WILLIAM WHISTON

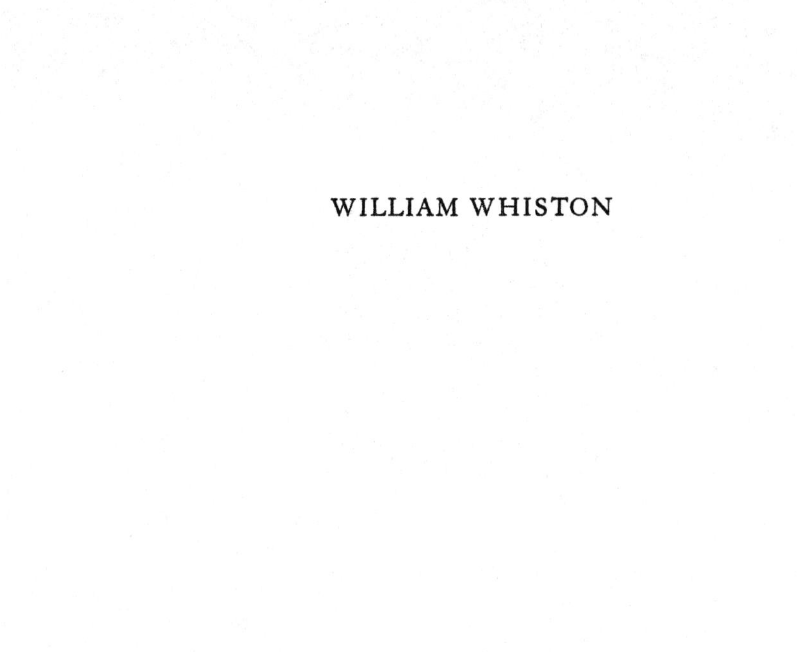

WILLIAM WHISTON

HONEST NEWTONIAN

JAMES E. FORCE
University of Kentucky

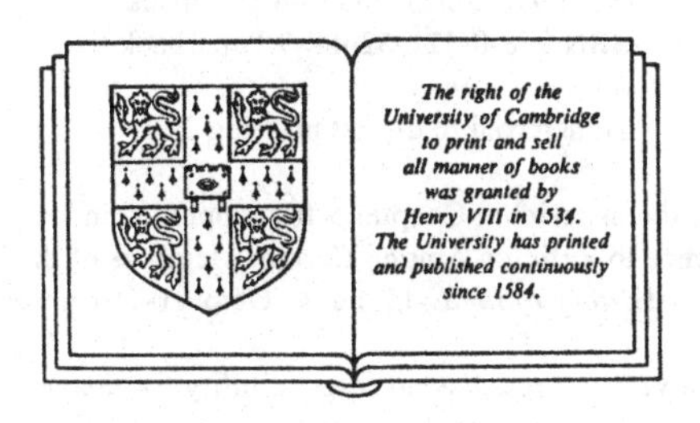

CAMBRIDGE UNIVERSITY PRESS
CAMBRIDGE
LONDON NEW YORK NEW ROCHELLE
MELBOURNE SYDNEY

CAMBRIDGE UNIVERSITY PRESS
Cambridge, New York, Melbourne, Madrid, Cape Town, Singapore,
São Paulo, Delhi, Dubai, Tokyo

Cambridge University Press
The Edinburgh Building, Cambridge CB2 8RU, UK

Published in the United States of America by Cambridge University Press, New York

www.cambridge.org
Information on this title: www.cambridge.org/9780521265904

First published 1985
First paperback edition 2002

A catalogue record for this publication is available from the British Library

Library of Congress Cataloguing in Publication data
Force, James E.
William Whiston, honest Newtonian.
Includes bibliographical references and index.
1. Whiston, William, 1667–1752. 2. Bible –
Prophecies – History – 17th century. 3. Millennialism –
History of doctrines – 17th century. I. Title.
BX5199.W52F67 1984 230′.092′4 84-4316

ISBN 978-0-521-26590-4 Hardback
ISBN 978-0-521-52488-9 Paperback

Transferred to digital printing 2009

Portions of the material in Chapter 5 first appeared in "Hume and
the Relation of Science to Religion Among Certain Members of the Royal Society,"
Journal of the History of Ideas 45, no. 4 (October–December 1984).

For Astrid

CONTENTS

ILLUSTRATIONS

FOREWORD

ALTHOUGH there have been a wealth of studies of millenarian theories and of prophets and prophecy over the last two decades, it is still a commonly held view that this sort of religious thinking ended when the Middle or Dark Ages ended. For those who work on the rich outburst of millenarian theories in England in the seventeenth century, it is often claimed that belief in these views ended with the death of Cromwell and the Restoration of Charles II. Before then, we are told, people believed that their world stood close to the end of time and hence took various worldly events, natural disasters, and prophecies as signs of what was coming. Further, it is often contended that the rise of the "new science" made reasonable, intellectual people give up such primitive beliefs and replace them with a world order regulated, maybe by God, in a strictly lawlike way. Newton's achievement was presumed to mark the end of the earlier millenarian view that rested on the prophecies set down principally in Daniel and the Revelation of Saint John. With Newton, the Enlightenment began.

This neat picture would be nice, as well as comforting to accept, were it not for some very odd facts. First of all, the great Sir Isaac Newton was a millenarian and wrote extensively on the prophecies in Daniel and Revelation. Second, his chosen successor as Lucasian Professor of Mathematics at Cambridge, William Whiston, was one of the most outspoken millenarians of the time and gave the public Boyle Lectures entitled *The Accomplishment of Scripture Prophecies*. Third, both Newton and Whiston were heirs to a continuous tradition, dating from Joseph Mede in the early seventeenth century, of seeing the growth of scientific knowledge as part of the fulfillment of the prophecies about the events leading up to the actual reign of Jesus Christ on earth for one thousand years. Fourth, the scientific-millenarian view continued after Newton and after Whiston in the writings of the founder of modern psychology, David Hartley, and those of the great physicist-chemist Joseph Priestley, who carried on this tradition up to the time of the French Revolution, which Priestley interpreted as the beginning of the end of this world.

These facts have been ignored, disregarded, and distorted to keep the image of our enlightened liberation through Newtonian science. "God said, 'Let Newton be, and all was light.'" Almost none of Newton's religious writings have been published. They have been treated as aberrant musings of a great scientific genius, having nothing serious to do with his monumental intellectual achievements. Whiston has been ignored or treated as a bad joke. The millenarian tradition has been divided into two parts: From Mede to 1660 it is recognized as a serious intellectual tradition, one that attracted the best minds of the day; from the Restoration onward it is considered to be a crackpot movement, outside of the world of scientifically oriented intellectuals. Hartley's religious writings are unread. Priestley wrote so much on religion that this aspect of his work cannot be ignored, but it can be treated in terms of his radical politics, his opposition to Hume, or his debates with the Scottish Common Sense philosophers. What is needed, and what Force's study of Whiston is part of, is a history of the millenarian-scientific idea from Mede to Priestley, and maybe beyond.

Joseph Mede, professor of Greek at Cambridge and the author of *Clavis Apocalyptica* (1627), was not a religious crank. He was a brilliant Cambridge student and teacher who had been reduced to complete skepticism by reading Sextus Empiricus in 1603 and who then tried to find an answer to his skeptical crisis in the intellectual disciplines taught at Cambridge. He found it only in the inspired reading of divine prophecies, especially in Revelation. A quiet, retiring scholar, he believed that the beginning of the millennium was imminent and that, as the Book of Daniel said, knowledge would increase as one approached the great event. It was now possible, as part of the growth of knowledge, to realize what the Book of Revelation was about and who the Antichrist was, as well as to understand more and more about the natural world. Mede's students and disciples, men such as Henry More, John Milton, Samuel Hartlib, John Dury, and William Twisse (the prolocutor of the Westminster Assembly), started preparing for the world to come by seeking to reform education and knowledge through planning new institutions and new theories. The outburst of intellectual energy and innovation produced by Mede's reading of the import of the prophecies is amazing. It was, of course, in some ways encouraged by the Puritan Revolution, in which all of Mede's disciples and students, except Henry More, played important parts. This side of the story has been told in depth by Charles Webster in his *Great Instauration*.

The question of what happened to millennium-oriented scientific activity after the Restoration, when the Royal Society was organized and chartered, is still being debated. Some of the most ardent millenarians were excluded. Samuel Hartlib, who had done so much to bring about organized scientific activity in preparation for the millennium, died before the Royal Society had to decide to exclude him because of his Puritan activities. John Dury, when he failed to interest Charles II in being a premillennial leader striving to convert the Jews, wisely stayed out of England for the rest of his life. However, even though the Royal Society made no overt effort to connect itself with its predecessor, the invisible college of the millennium-oriented scientists, it did have some of the same members: Robert Boyle, John Wilkins, and Henry Oldenburg. Boyle was Dury's nephew and Oldenburg his son-in-law. The Oldenburg–Boyle correspondence during the Jewish messianic movement of 1666 shows that these two, at least, retained interest in signs of the coming of the millennium. They also stayed in contact with leading Dutch millenarians.

By the time Newton appeared on the scene, he was concerned both with the scientific problems being dealt with by the Society and with the interpretations of Daniel and Revelation being discussed by his teacher and friend Henry More. During Newton's most productive period as a scientist, he was also working continuously with More in unraveling the prophecies in Scripture, and More was telling him of the latest treasures he had found in the cabalistic manuscripts coming from Palestine. Newton, as we know, long delayed publishing the principle of universal gravitation because he was so involved with More in millennial research. More's description indicates how involved Newton was in this.

Whiston first became known through his work, *The New Theory of the Earth,* published in 1696. In it he used the principles of Newtonian physics to show how the earth began, survived the Flood, and would perish–all in keeping with the descriptions in the Bible. Newtonian physics became the scientific explanation of how the scriptural world started, persevered through the great cataclysm of the Flood, and would be consumed, as described in the prophecies in the Book of Revelation. As Force shows, Newton was pleased with this union of the scriptural account of God's way of running the world and the physical system presented in the *Principia,* and when he retired from the Lucasian Chair of Mathematics, he had young Whiston appointed to it.

Whiston, as a scientific theologian, also evaluated the Christian prophecies in terms of the probabilistic theory of knowledge being developed by the latitudinarian divines in the Royal Society, the legal theorists, and the early empirical philosophers of science (described in Henry G. van Leeuwen's *Problem of Certainty in English Thought*). Using the theory of the degrees of certainty and of restricting evidence to that available for a given problem, Whiston tried in his Boyle Lectures to present an empirical probabilistic case establishing that a large number of biblical prophecies had been fulfilled within biblical history and others in postbiblical history. Therefore a reasonable man should expect the rest of the prophecies, those related to the occurrence of the millennium, to be fulfilled. Thus the reasonable, scientifically oriented person should expect the conversion of the Jews, their restoration in Israel, the overthrow of Antichrist, the rebuilding of the Temple, and the Second Coming.

Whiston's argument shows that millennial thinking was alive and well in the best intellectual circles in England in the first decade of the eighteenth century. Whiston's subsequent career as public heretic and bad prophet may have made him the butt of all sorts of jokes during the second, third, and fourth decades of the century, but he started out as Newton's personally chosen successor and spokesman, probably because of the way he combined Newtonian physics and scriptural prophecies about the millennium.

Almost ten years ago, Professor Force, then a graduate student, my colleague Richard A. Watson, and myself proposed a grandiose project to the National Science Foundation: to explain the role of millenarian thought in the development of modern philosophy. We were given enough support to start on this. And now I think we each, from our own angle, see that millenarian theorizing was a most important element in the course of "the making of the modern mind" that can only be ignored at one's peril. To try to appreciate the enormous intellectual journey that Western man made between 1600 and 1800 without taking account of the rise and transformation of millenarianism is to miss some of the crucial dynamics of what happened – and to misunderstand where we are today and how we got here.

Millennial ideas played a vital role in delineating the roles that various national states thought they were playing during this period. The political developments in Western and central Europe – from Spain and Portugal to England and Scotland, France, Sweden, Poland, and various Protestant and Catholic states in Germany, Austria, and Bohemia – reflect, at least in part, the role that rulers and church leaders believed their countries and their particular churches would play in the transformation of the world. The Reformation and Counter-Reformation, the conversion of most of the Spanish and Portuguese Jews, the discovery of the New World, the rediscovery of antiquity, the spreading of European culture all over the globe – all reinforced the conviction of many forward-looking philosophers, scientists, and theologians that some monumental change was taking place. The identification of these events with the scenario in the books of Revelation and Daniel, and with the role that the Jews were expected to play, was a major feature in the thought of certain sixteenth-century theologians, from the Protestants John Napier, Thomas Brightman, and John Henry Alsted to Catholic theologians such as Luis de Leon and Guillaume Postel. Two kinds of seventeenth-century millenarian speculation seem to have inspired major intellectual developments. The conviction of Mede, Twisse, Comenius, Dury, and Hartlib that the increase of knowledge was to play a crucial part in preparing for the millennium led to all sorts of proposals and actual programs for reforming the content and institutions of learning and knowledge. One result, modern education, fostered by Comenius's textbooks, new curricula, and the introduction of scientific research, developed from a particular way some intellectuals tried to deal with the skeptical crisis of the late sixteenth century by finding certainty in what they considered to be divinely inspired readings of biblical prophecies. The Cambridge Platonists, mainly students of Joseph Mede, tried to provide a philosophy to understand this outlook. The metaphysics they developed led to Isaac Newton's view of the world. The epistemology of Henry More played an important role in formulating the theory of scientific knowledge of the Royal Society.

Another kind of millenarian speculation posed serious questions about the accuracy of existing scriptural texts and of Scripture as a picture of world history. A peculiar French millenarian, Isaac La Peyrère, published two works, *Du rappel des Juifs,* 1643, and *Prae-Adamitae* (*Men before Adam*), 1655. La Peyrère, who was a secretary to the prince of Condé and a French envoy to the court of Queen Christina, was part of the avant-garde intellectual circle in Paris that included Gassendi, Mersenne, Grotius, and Hobbes, among others. La Peyrère's millenarian message was that the Jews would soon be recalled to the center of world history and that they, led by the king of France, would rebuild the Holy Land, from which the Jewish Messiah would emerge to rule the world – along with his regent, the king of France. As part of the justification of this message, La Peyrère insisted that the existing text of the Bible was not accurate, that the Bible was not the history of the world but only the history of the Jews, that Moses was not the author of the Pentateuch, and that there were men before Adam, who were not part of the creation drama described by the Bible. The pre-Adamites were the ancestors of the many different kinds of human beings being discovered all over the planet. La Peyrère used a battery of internal and external points about the scriptural text to buttress his case. Much of it was immediately taken over by the budding "higher criticism" of the Bible. Spinoza, Richard Simon, and other brave souls took over some of the challenges to the Bible's accuracy and relevance and developed a skepticism about some of the aspects of traditional religion.

Principally in England, liberal Anglican theologians in the seventeenth century sought to develop a moderate, commonsense answer to the skepticism emerging from Descartes's work and from La Peyrère's critique of the Bible, and also to the Arianism (the denial of the doctrine of the Trinity and often of Jesus' divinity) that was accepted by many millenarians. Latitudinarism seems to have grown out of the skeptical epistemological attacks on Protestantism that were launched by French and English Jesuits; out of claims to unique, exclusive religious knowledge made by various kinds of Dissenters, independents, Levelers, Ranters, and Quakers; and out of skeptical challenges to the content and accuracy of Scripture. A kind of mitigated skepticism was developed first by William Chillingworth, in answer to the Jesuits. This view was worked out in more systematic detail by Henry More in his *Antidote to Atheism* and his answer to the enthusiasts. More introduced a most radical scepticism, arguing that everything we know or believe could be false or dubious if our faculties were deluded or delusive. If one moderates one's quest for certainty to a reasonable level, however, one then sees that the evidence for More's version of Christianity is convincing–although it could still be false. We have no infallible certainty, but we have adequate or "indubitable" certainty, and this is the most that we can attain in mathematics, religion, and science.

More's skepticism and its cure were adopted by latitudinarians such as Stillingfleet, Wilkins, and Tillotson as a way of rebutting the emerging religious skepticism stemming from the biblical criticism of La Peyrère, Simon, and Spinoza. The latitudinarians sought to show that it was reasonable to accept the text of the Bible as accurate and to accept its account of the early history of the world. One could not answer or overcome all possible questions or doubts on these matters, but a reasonable person could easily see that there was at least as much evidence for accepting the Bible as for accepting any historical document. Considering how important the Bible is and has been, they argued, its text could not have been altered unless there were a general conspiracy to mislead mankind, a most unlikely possibility.

If we can have reasonable assurance that the text is accurate, we can then also be reasonably certain, though we cannot prove with mathematical certainty, that Moses wrote the Pentateuch and that the Apostles wrote the Gospels attributed to them. We have no reasonable basis for suspecting that Moses or the Apostles had any desire to deceive us. So, we can reject La Peyrère and Spinoza and trust the content of the Mosaic account and the New Testament. There is no reason to suspect there were men before Adam, that Noah's Flood was not universal, or that the messiah came in the year 1 (A.D.). We have as much reason to believe in a miracle reported in the Bible as we have in any other matter of fact. The reasonable, scientifically minded person will see how unreasonable the developing irreligious skepticism is.

In his particular version of this "reasonable" defense of religion, Whiston developed a "reasonable" basis for expecting the fulfillment of the scriptural prophecies. In his Boyle Lectures, he used the latitudinarian argument concerning reasonable or moral certainty to defend the truth of scriptural prophecies. Anyone who accepts the Bible as history, he said, can tell that many prophecies recorded in the Old Testament were fulfilled later on in either Jewish or Christian history. In addition, many prophecies in the New Testament have been fulfilled in postbiblical history. Therefore, when one considers how many prophecies have been fulfilled already, one would have to be an arrant skeptic to doubt that others will not be fulfilled. Whiston tried to put acceptance of prophecy fulfillment beyond reasonable doubt. And, in so doing, he marked out a defense of our ability to have knowledge of the future that was expanded upon by Bishop Clayton, Bishop Newton, and many others in the eighteenth century.

Whiston was one of many seeking signs of the fulfillment of the greatest prophecies of all, those predicting the coming of the millennium and the end of the world. He may have become a comic character as he saw every earth tremor as a possible sign, but Whiston was part of a large and serious movement of intellectuals in Europe and America who had "reasonable" expectations that the grand scenarios of the books of Daniel and Revelation would be played out in the eighteenth century. He died before the Lisbon earthquake and before the momentous events that were to lead to a climax of reasonable, antiskeptical interpretation of actual history as the fulfillment of biblical prophecies – namely the American and French revolutions. The latter contained such dramatic developments as the dethroning of the church, the capture of the pope, the overthrowing of the kings of this world, the establishment of a new world empire, the recall of the Jews, at last, as citizens in America and France, and the establishment of the Paris Sanhedrin. The outpouring of millenarian interpretation of the two revolutions and the Napoleonic age is immense, much of it long forgotten and unstudied. Occasionally some of the more bizarre figures and readings are revived to show us how unenlightened and gullible some people were. But we are rarely reminded of how much serious, "scientific," "reasonable" millenarian literature there was: Joseph Priestley, interpreting the French Revolution from the vantage points of England and America as the beginning of the glorious finale of human history; Abbé Henri Grégoire, one of the first leaders of the French Revolution, seeing it as the onset of the millennium; the American revolutionary leader Elias Boudinot, writing *The Age of Revelation* to answer Tom Paine's *Age of Reason* and resigning as first director of the U.S. Mint to write *The Second Advent.* As Grégoire and Boudinot discerned, Napoleon was using this wave of millenarian excitement to cast himself as the central character, the secular messiah, reviving La Peyrère's claim that the king of France would rule the world with the messiah.

The aftermath of the Napoleonic era seems to have left a large gulf between those who still expected the scenario of Daniel and Revelation to occur on earth and those who sought new secular explanations of world developments in terms of nationalism or socialism. Some, like the great Polish writer Mickewicz, developed a nationalist millenarianism – Poland was the Christ of nations, and the climax of history would occur with the resurrection of the crucified Polish nation. Others saw the world drama in terms of class struggle, the growth of the British Empire, the unification of Italy or Germany, or America's Manifest Destiny. These secularizations of millenarianism pushed the purely biblical millenarians off center stage into fundamentalist or evangelical groups. The catastrophe of World War II and the emergence of the state of Israel has now made these groups more vocal, and more serious continuers of the seventeenth- and eighteenth-century tradition who foresee Armageddon in the form of a nuclear holocaust and the Second Coming of the Messiah as the expected development from Israel's history.

Millenarianism was a vital intellectual and spiritual force, which enjoyed great intellectual respectability up to the end of Napoleon's reign. It is rising again, but without its great scientific and theological spokesmen. To appreciate one of its high points is to see it as the union of the new science and a defense against religious skepticism. Whiston was one of the major figures of scientific millenarianism. This study of him should make us realize that serious, intellectualized millenarianism outlasted the Restoration and established a framework within which serious, scientifically oriented intellectuals could interpret the great political upheavals at the end of the eighteenth century. The good – or sad – fact that the United States and the French Republic became secular states with secular destinies has blinded us to the force of a major way of understanding events. We have traced the major line of development in European intellectual history from Erasmus and Montaigne to Bacon, Descartes, Spinoza, and the French Enlightenment. In constructing such a lineage, we have lost track of one of the most vigorous sets of ideas that influenced the world view of Europeans. We have so lost track of this that we think of the millenarian reading of history as some kind of oddball, crankish, irrelevant view and are surprised to find it popping up in statements by U.S. cabinet ministers, a president of Guatemala, and various Jewish and Christian spokesmen in America and Israel. To recapture the millenarian dimension of our intellectual history, we need studies such as this one by Professor Force on Whiston, which helps us to see how religion, science, and philosophy interacted in forming our conception of what kind of a world we are living in and where it is going. With their aid we will be able to see that millenarian thinking has been, and still is, a major way of reading events, one that though changing and developing, has helped to share men's thinking from the Reformation to the latest pronouncements of the Rev. Jerry Falwell.

RICHARD H. POPKIN
Washington University

PREFACE

My interest in the thought of William Whiston began in 1975, when, as a graduate student at Washington University in St. Louis, I became engaged in an extensive study of Hume's *Dialogues Concerning Natural Religion.* In reading Richard H. Hurlbutt's incisive study *Hume, Newton, and the Design Argument* (Lincoln: University of Nebraska Press, 1965), I was impressed by Hurlbutt's persuasive argument that Hume's critique of natural theology and the design argument "was aimed primarily at the Newtonian attempt to bring about a rapprochement between science and religion" (p. xii).

I became further absorbed in the question of the "rapprochement" between Newtonian science and religion when I attended Richard H. Popkin's Washington University seminar on millennialism in the spring of 1975. In that class Popkin emphasized the role of the interpretation of millennial prophecy and biblical criticism in the development of the historical social sciences. In the course of my readings for this class, I encountered Frank E. Manuel's book *Isaac Newton, Historian* (Cambridge, Mass.: Harvard University Press, 1963). Manuel, in the following vivid passage, brought to my attention a man who seemed to be a key figure in the development of a specifically Newtonian rapprochement of science and religion:

> Of all the members of the Newton circle William Whiston, his successor in the Lucasian chair at Cambridge, was far the most prolific writer on Mosaic astronomy, on ancient chronology, on the interpretation of the Old and the New Testament, on problems of the Trinity, on the coming of the great conflagration, and on prophecy – he compiled more than 40 works. His eccentric character led him to exaggerate where the more respectable figures in Newton's group either kept their opinions secret or restrained their enthusiasm. Yet precisely because they are so forthright, William Whiston's works cast important light on the hidden intent and meaning of similar writings by his great contemporary. Where Newton was covert Whiston shrieked in the marketplace. [p. 143]

This quotation delighted me because it suggested that the Newtonian rapprochement between science and religion was more wide-ranging than the design argument that had been the focus of Hurlbutt's analysis. Popkin's emphasis on biblical criticism and prophecy interpretation in the seminar seemed to find its paradigmatic Newtonian expression in the thought of William Whiston, and I immediately set about gathering what information about him was available to me in St. Louis. I hoped to write my doctoral dissertation on this fascinating figure. This hope was nearly dashed when I learned that Sister Maureen Farrell had already written a thesis on Whiston at the University of Manchester in 1973.

An examination of a microfilm of her thesis, published in 1981 by Arno Press in their series "The Development of Science," revealed precise and detailed scholarship. I felt, however, that her biobibliographical focus on Whiston's contributions to the science of his day differed sufficiently from my own concern with the nature of the Newtonian rapprochement of science and religion as reflected by Whiston to enable me to continue my project.

From 1975 to 1977 I worked on this project in London under the auspices of grants from the National Science Foundation and the National Endowment for the Humanities and eventually produced, in the fall of 1977, after returning home again to Washington University in St. Louis, a preliminary investigation of the subject in the form of a dissertation entitled "Whiston Controversies: The Development of 'Newtonianism' in the Thought of William Whiston."

I set this entire project aside for four years while teaching at the University of Kentucky. In the spring of 1982, I was given a chance to take the spring semester off from Kentucky and to use the resources of the William Andrews Clark Memorial Library at the University of California at Los Angeles to complete my work on Whiston, to prune and revise my initial essay by entering upon the daunting terrain of Whiston's vast output of varied works. The present book, with 30 percent new material developed from my research at the Clark Library, is the result. Here I attempt to focus on Whiston's uniquely Newtonian synthesis (or rapprochement) of the design argument of natural religion with revealed religion. I have attempted in this work to place Whiston's work in the social and historical context of his own times and also to take into account the results of my analysis of Whiston for the single most illuminating work on Newtonianism in the past several decades, Margaret C. Jacob's *Newtonians and the English Revolution, 1689–1720* (Ithaca: Cornell University Press, 1976). In the summer of 1983 I returned to the Clark Library and made revisions in the manuscript.

Given the length of time that I have been engaged with this project, almost a decade, it is impossible to acknowledge my gratitude to all of my friends, colleagues, and teachers who have contributed to its completion. Nevertheless, thanks are owed, first and foremost, to Professor Richard H. Popkin of Washington University in St. Louis for his scholarly inspiration and friendly encouragement over the years. Professor Richard S. Westfall of Indiana University very graciously read a draft version of the manuscript and made many valuable criticisms, as did Professor Richard G. Olson of Harvey Mudd College in Claremont, California. I am also indebted in a variety of ways to Professors Richard A. Watson and Jerome Schiller of Washington University and to Professor Charles Schmitt of the Warburg Institute in London. In addition, I should like to thank the staff of the incomparable William Andrews Clark Memorial Library, especially Professor Norman J. W. Thrower, Director; Tom Wright, Clark Librarian; John Bidwell, Assistant Librarian; and the most helpful group of Librarians in the world, especially Carol Briggs, Patrick McCloskey, Ray Reece, and Carol Sommer.

Material support over the last decade has been generously provided by the National Endowment for the Humanities; the National Science Foundation (where Dr. Ronald J. Overmann, director of the History and Philosophy of Science Program, has been particularly encouraging and helpful); the University of Kentucky Faculty Research Grant Committee; the University of Kentucky Summer Research Fellowship Committee; and the William Andrews Clark Memorial Library Fellowship Committee.

The entire work was typed by Pat Harris, Sylvia Henderson, Linda Bowers, of the Department of Philosophy at the University of Kentucky, and by Pat Willhite of the university's Honors Program, and I am extremely grateful to them all.

A final word of thanks is owed to Dr. Christie Lerch for her really superb job of copy-editing and to Margo Shearman-Howard, who expertly read the proofs and provided the index.

Anybody who has ever undertaken a similar project appreciates how indispensable is the unstinting effort of such dedicated people.

INTRODUCTION: *ERKENNEN* AND *VERSTEHEN* IN THE HISTORY OF PHILOSOPHY, AND WILLIAM WHISTON'S "NEWTONIANISM"

In the Introduction to his ***Philosophy as Social Expression,*** Albert William Levi contrasts in detail two theoretical schools of the history of philosophy.[1] According to one school of theorists, the philosophically important aspect of a text is the text itself, which, it is maintained, is logically independent of, and intellectually autonomous from, any historical context. All that is relevant to the understanding of any philosophical text is carried timelessly in the text itself. Levi often refers to this school as semantic "atomism."

Contrasting to this assumption of a permanence of meaning that is outside of time, locked ahistorically in "atoms" of text, is an opposing theoretical school according to which terms and arguments in the history of philosophy must be interpreted within the special framework of concepts and distinctions specific to the thinker's cultural context. The history of philosophy, for this "contextualist" school, focuses on how a philosopher's works respond to the challenges of a particular historical milieu and the questions inherent within a specific social structure.

Contextualists assert that the atomistic doctrine of textual autonomy produces only a "mere understanding" (*Verstehen*) of philosophical works. "Full comprehension" (*Erkennen*), on the other hand, depends on contextual analysis. Contextualists argue that for a "fuller comprehension" of any philosophical work one must examine biographical details of the life of the writer, the predecessors to whom he is intellectually indebted and successors whom he influenced, how his works fit together and into the cultural ideals of the era, and the author's class affiliation and place in the social hierarchy of the age. Levi argues that "these questions are *not* philosophically irrelevant. They are the very vehicles which lead us from the lean subsistence of 'mere understanding' *toward* the more generous nourishment of a 'fuller comprehension' of the great philosophers and their classic texts."[2]

Levi notes that in the actual practice of writing the history of philosophy, both strict analysis of the text as a logically autonomous, semantically meaningful atom and analysis of it as an expression of a particular philosopher's individual makeup and social context are required but that the most important level of analysis, the level leading toward a "fuller comprehension" of philosophical writing, is the contextual.

I agree with Levi's approach. The present work is an attempt to come to both an understanding and a fuller comprehension of the rapprochement between science and religion in the Newtonian context by focusing on the controversies concerning William Whiston. Frank Manuel is correct when he says that whereas Newton was cautious about expressing his heterodox religious views, his forthright disciple Whiston shrieked them out in the marketplace.[3] More recently, an eminent biographer of Newton's, Richard S. Westfall, suggested that in Whiston's memoirs "one catches a glimpse – is it a true image or is it a mirage? – of one of the most advanced circles of free thought in England grouped about Newton and taking its inspiration from him."[4]

Despite these suggestions of Whiston's importance for understanding the Newtonian rapprochement between science and religion, Whiston's work has not yet received a full-length analysis that focuses on the light it sheds on the concealed religious views of his great contemporary Isaac Newton. Maureen Farrell is the only writer who focuses exclusively on Whiston's life and his contributions to such scientific subjects as cosmology, longitude, mathematics, and astronomy. Farrell does devote one forty-seven–page chapter to a summary of Whiston's religious thought, but in general her theme is that Whiston "maintained concurrently an active interest as much in scientific as in religious studies."[5] Undoubtedly her substantial and important contribution in documenting the biographical details of Whiston's life and scientific achievements marked a signal advance in Whiston studies. Most other scholars have mentioned Whiston only in passing and in connection with their own central concerns. Thus Robert H. Hurlbutt briefly treats Whiston's essentially Newtonian version of the design argument.[6] Hélène Metzger similarly discusses Whiston's analysis of gravity as inessential to bodies within the context of the Newtonian design argument.[7] Many scholars, most notably D. C. Kubrin, have traced Whiston's views on the creation and dissolution of the world (see Chapter 2, note 4). Frank E. Manuel has touched upon Whiston's views regarding biblical chronology in relation to those of Newton.[8] Richard S. Westfall, in his prize-winning biography of Newton, has provided a significant sketch of Whiston's relationship with Newton.[9] Finally, Margaret C. Jacob has indicated Whiston's significance in the Newtonian social and political program. Of all these scholars, Jacob is the one most interested in tracing the impact of Newton's many-sided genius in the social and political life of the time as it is revealed in the thought of his followers.[10] By studying the Newtonians who expounded, extrapolated, developed, and promulgated Newton's ideas in the sociopolitical arena, she indeed moves toward a fuller comprehension of Newtonianism.

Just as Jacob attempts to widen the meaning of "Newtonianism" by examining this largely ignored sociopolitical aspect of Newton's impact on his own era, I, too, attempt to show how Newtonianism in the social, political, and theological arenas includes much more than the overfamiliar Newtonian version of the design argument. What Jacob does with the Newtonian movement as a whole, I attempt to do in microcosm by examining in detail the writings of one of the most important but least studied members of the Newtonian circle, William Whiston, within the context of the intellectual concerns eddying around the Newtonian movement at the turn of the seventeenth century. Many of the most controversial social, theological, and political aspects of Whiston's theory of the relationship of science and religion were also subscribed to by Newton.

In Chapter 1, I trace the temper and times of the Newtonian controversialist who served, in all probability, as the model for that eighteenth-century caricature of comic integrity, the Rev. Dr. Primrose in Oliver Goldsmith's *Vicar of Wakefield*. Relying primarily on Whiston's own accounts of his life and times, I seek to establish the sort of society in which Whiston lived and for which he wrote. Whiston's autobiography also reveals the temperament of an extraordinary individual bent on following his own "Primrose" path by hewing to religious principles that he felt to be an extremely important corollary to Newton's natural philosophy. Newton apparently agreed, and took an active part in bringing Whiston back to Cambridge from his country vicarage, first as Newton's substitute and then as his successor in the Lucasian Chair of Mathematics. In his own age, Whiston's eccentricity appeared to turn on his self-martyrdom over the touchy political and religious issue of Arianism, or antitrinitarianism. Because he honestly stated his religious convictions, Whiston lost his job at Cambridge in 1710. If his contemporaries found him overeager to sacrifice his career for the sake of principle, Whiston found his contemporaries far too ready to sacrifice any and all principle for the sake of preferment. When he once chided Sir Richard Steele for knuckling under to government pressure by speaking in the House in favor of the directors of the South Sea Bubble just after he had denounced them in his newspaper, Steele replied, "Mr. Whiston, you can walk on foot [do without a carriage], and I cannot."[11]

In Chapters 2 and 3 I examine the vocabulary of terms and distinctions that Whiston used in formulating his theories and the level of certainty that he believed it was possible to attain with his methodology. The upshot of these two chapters is to widen considerably the meaning of the term *Newtonianism*.

Chapter 2 includes an analysis of Whiston's *New Theory of the Earth* (1696), a theory of the origin of the earth that is also a detailed refutation of Thomas Burnet's earlier *Sacred Theory of the Earth*. Whiston's methodological approach to geocosmology is introduced by a separately paginated ninety-five–page introduction entitled "A Large Introductory Discourse Concerning the genuine Nature, Stile, and Extent of the Mosaick History of the Creation." In this introduction to his *New Theory,* Whiston aims to correct Burnet's assertion that the Mosaic account of creation is a "meer Popular, Parabolick, or Mythological relation," in a fashion reminiscent of Newton's own methodology regarding the proper method of interpreting Genesis that Newton had outlined to Burnet in a series of letters in 1681.

Newton's letters to Burnet and Whiston's introduction to his *New Theory* show their agreement that the scriptural narrative of creation, although historically accurate in a sense, is not a "Nice and Philosophical Account of the Origin of All Things." As Whiston echoes and elaborates Newton's remarks to Burnet, it becomes evident that what would count for both as a "Nice and Philosophical Account" of the creation of the world would be a "mechanical" description, consistent with natural law, of the secondary causes in the natural world that occasioned the gradual transformations of the world that occurred in the period from the first moment of creation through the time of the Flood. Whiston strives mightily to thread his way between the "wildness and unreasonableness" of simple-mindedly literal interpreters and those such as Burnet who provide ammunition for deistic ridicule of revelation by asserting the Bible to be a "mythological relation." For Whiston, the thread of Ariadne in providing a "Nice and Philosophical Account" of the historically accurate Mosaic narrative is the sure and certain natural philosophy of his great mentor, Isaac Newton. Whiston says:

> Since it has now pleased God, as we have seen, to discover many noble and important Truths to us, by the Light of Nature, and the System of the World; as also, he has long discovered many more noble and important Truths by Revelation, in the Sacred Books; It cannot be now improper, to compare these two Divine Volumes, as I may well call them, together; in such Cases, I mean, of Revelation, as relate to the Natural World, and wherein we may be assisted the better to judge, by the Knowledge of the System of the Universe about us. For if those things contained in Scripture be true, and really deriv'd from the Author of Nature, we

> shall find them, in proper Cases, confirm'd by the System of the World; and the Frame of Nature will in some Degree, bear Witness to the Revelation.[12]

There is no evidence that Whiston saw Newton's letters to Burnet or that he was even aware of them. Nevertheless, I believe that I am justified in arguing, as I do in Chapter 2, that Whiston's method of interpreting Genesis is Newtonian in a strong sense. Without doubt, if the argument that Whiston represents or follows Newton's own method were based solely on the inner consistency of these two approaches to reconciling Scripture with science, that would be a long way from establishing any specifically Newtonian connection between them. Furthermore, one of their basic points of agreement – the idea that Genesis is an account of what an observer present at the creation would literally have seen – is a commonplace throughout the seventeenth and eighteenth centuries. For example, when Lady Percival queried George Berkeley about the Mosaic account of creation within the context of the last sections of Berkeley's *Treatise Concerning the Principles of Human Knowledge,* Berkeley answered that "to agree with the Mosaic account of the creation it is sufficient if we suppose that a man, in case he was then created and existing at the time of the chaos, might have perceived all things formed out of it in the very order set down in Scripture which is in no ways repugnant to our principles."[13] In short, if there were no other evidence of a connection between Whiston's and Newton's methods than that basic similarity, the most I could claim is that Whiston's method is Newtonian only in the sense of its being shared by Newton and many other people as well.

My argument that Whiston's views of interpreting Genesis – and, indeed, Scripture as a whole – reflect Newton's views is based on historical evidence. The well-documented fact that Newton played an active role in bringing Whiston (his former pupil) back to teach at Cambridge, first as Newton's own substitute, with the full profits of the Lucasian Chair, and then as his successor, seems at least to suggest that Newton shared Whiston's views. We also have Whiston's statement that his *New Theory of the Earth* was "chiefly laid before Sir *Isaac Newton* himself, on whose Principles it depended, and who well approved of it."[14] It is quite probable that the "principles" Whiston is referring to are his principles ("postulata") for interpreting Scripture, which are listed at the end of the long "Introductory Discourse" on "the Mosaick History of the Creation." The first two of these postulates had been expressed in germ in the letters to Burnet by Newton fifteen years before. In Whiston's hands they become the basis for his interpretation of fulfilled historical prophecies and unfulfilled future prophecies of the Apocalypse, and also the basis for his radical Arianism and Whiggish theorizing. All of Whiston's strident controversies are rooted in these "postulata":

> I. The Obvious or Literal Sense of Scripture is the True and Real one, where no evident Reason can be given to the contrary.
> II. That which is clearly accountable in a natural way, is not without reason to be ascrib'd to a Miraculous Power.
> III. What Ancient Tradition asserts of the constitution of Nature, or of the Origin and Primitive States of the World, is to be allow'd for True, where 'tis fully agreeable to Scripture, Reason, and Philosophy.[15]

The historical thesis that Newton shared at least two of these principles of biblical interpretation, which go well beyond the common conviction that the divine volumes of nature and Scripture are harmonious, is further strengthened in Chapter 3, where I reveal evidence that Newton continued to act behind the scenes to promote Whiston's career by suggesting the topic of Whiston's 1707 Boyle Lectures, *The Accomplishment of Scripture Prophecy*. The proper interpretation of fulfilled historical prophecies such as those regarding the messiahship of Jesus may seem to be a different problem from that of interpreting the nature of the prophetic history of Genesis in a manner harmonious to science, but in fact for Whiston, and probably for Newton, it was not. Both depend on the postulates, especially Postulate I, according to which "the Obvious or Literal Sense of Scripture is the True and Real one, where no evident Reason can be given to the contrary." The prophetic language of Moses is a literal prediction of determinate historical events. So, too, the language of later biblical prophets uttering their predictions of the future may appear "peculiar and enigmatical" in style, but each prophetic prediction nevertheless points to one, and only one, determinate historical event as its fulfillment. To interpret prophecies in terms of allegorical double meanings is unreasonable, as Whiston argues in his Boyle Lectures:

> If Prophecies are allow'd to have more than one event in view at the same time, we can never be satisfy'd but they may have as many as any Visionary pleases; and so instead of being capable of a direct and plain Exposition to the satisfaction of the judicious, will be still liable to foolish application of fanciful and enthusiastick Men.[16]

Because this entire approach to prophecy interpretation was apparently suggested to Whiston by Newton; and because it is based on Postulates I and II from Whiston's introduction to his *New Theory,* which Whiston claimed that Newton "well approved"; and because of Newton's subsequent and decisive intervention in Whiston's academic career, I believe that I am justified in my claim that these principles of interpretation were shared by Newton and Whiston.

Chapter 3 is designed to buttress my thesis that Newtonianism properly contains an overlooked facet – biblical interpretation – and that William Whiston's many works illustrate this neglected facet of Newtonianism. At the conclusion of the chapter I trace the implications of this thesis for the work of Margaret Jacob. Jacob's main point is that young Newtonian scientist-theologians such as Richard Bentley, Samuel Clarke, and William Whiston united with moderate, Low Church bishops such as William Lloyd, Simon Patrick, and William Wake to adapt the Newtonian model of the universe – which conceives of it as being designed by a generally provident grand architect greatly skilled in mathematics – as a model for society, following the civil and religious chaos of the Glorious Revolution.[17] As far as she goes, Jacob is correct. The design argument is certainly emphasized as a model of order and stability for social purposes by Whiston, who sees the deists, with their mockery of Scripture, as a great threat to the moral fabric of society. I seek to supplement Jacob's thesis, however, by showing how much more Newtonianism encompasses than the design argument. It contains a whole program of biblical interpretation and criticism, part of which uses the design argument to confirm the verisimilitude of Scripture. Second, Jacob's assertion that the Boyle Lectures were a primary platform for the public dissemination of this more widely focused Newtonianism is buttressed by Whiston's claim that his own Boyle Lectures on *The Accomplishment of Scripture Prophecy* were originally suggested to him by Newton.

In Chapter 4, I examine Whiston's application of his Newtonian method of biblical interpretation to three raging controversies of the time, concerning the nature of royal authority, antitrinitarianism, and millennialism. Whiston argues, in his *Scripture Politicks,* that a king's right to rule is bestowed providentially by God through the mechanism of the choice and recognition of the people. Whiston also applies his Newtonian historical method of textual interpretation to the development of an antitrinitarian theology, arguing that the doctrine of the Trinity is a cruel hoax perpetrated by Athanasius. The Clark Library at the University of California at Los Angeles possesses a variant of one of Newton's manuscripts detailing the history of the church in the first centuries. This manuscript, and Whiston's book entitled simply *Athanasius Convicted of Forgery,* document that Newton and Whiston shared the view that Athanasius was a forger. Finally, I trace Whiston's millennial expectations, which arise from his Newtonian method of biblical interpretation. For Whiston, as well as for Newton, contractarian (and providential) Whig political theory and reformed church doctrine that culminate in antitrinitarianism and the expectation of an apocalyptic Second Coming of the Messiah are all connected by Whiston's distinctively Newtonian method of biblical interpretation, even though Whiston and Newton differed significantly regarding the imminence of the Apocalypse.

Whiston's use of the Newtonian method of scriptural exegesis forces a modification of Jacob's view that by 1720 Newtonianism had triumphed as a social philosophy. For Jacob, Newtonianism as a social and political force means the Newtonian design argument with its stable universe, providentially designed by God, operating by natural laws as the model for church and state. By identifying Newtonianism exclusively with the design argument, Jacob ignores the specifically Newtonian scriptural basis for legitimizing the Glorious Revolution, Arianism, and millennialism. Jacob's sanitized, design-oriented, social Newtonianism does emerge, with some qualifications, as triumphant in the social and political arenas by 1720, but this version of Newtonianism is not the entire story. When the excised portions of the Newtonian social, theological, and political program are restored, Newtonianism emerges as a richer, more complex social and political force in the political context of the first decades of the eighteenth century. However, it cannot in any way be considered triumphant. After 1710, Whiston's application of his Newtonian method of biblical interpretation to such theologically sensitive issues as the doctrine of the Trinity and the imminence of the Second Coming of the Messiah led to the collapse of Whiston's academic career and the beginning of his reputation as a learned crackpot. Newton retreated further into his zone of silence, trusting that the wise would understand. The fact that so astute an interpreter as Jacob can identify the sociopolitical program of Newtonianism entirely with the design argument, leaving entirely out of account the Newtonian scriptural interpretations that profoundly affected society and politics, shows just how unsuccessful this aspect of Newtonianism was. Even by the middle of the eighteenth century, however, this aspect of Newtonianism had been eliminated.

In the preceding chapters, I argue that Whiston shares a distinctive basic approach to interpreting Scripture with Newton and that he elaborates this Newtonian approach to Mosaic history, fulfilled historical prophecies, the created and hence inferior nature of Jesus in comparison to God the Father, and the coming Apocalypse of unfulfilled future prophecies in a manner generally in accord with Newton's private thoughts on these subjects. In Chapter 5, I seek to show their most basic disagreement about the Bible by clarifying their respective positions with regard to deism. Many of Whiston's controversies, such as that concerning the proper method of interpreting Genesis and fulfilled prophecies, grow out of his opposition to such prominent deists as Charles Blount and Anthony Collins. Whiston strives always to prevent the mocking spirit of such men from demoting the primary status of the revealed word even while they accept the God of the design argument. Whiston's work, intended to illustrate through the design argument the generally provident architect-creator God and through his analysis of biblical prophecies the continuing, specially provident, miracle-working, prophecy-fulfilling God of revelation (properly interpreted), is aimed primarily at the deists and also fits into the wider context of similar efforts by certain members of the early Royal Society. Whiston is always guided in his controversies with the deists by his third postulate: "What Ancient Tradition asserts of the constitution of Nature, or of the Origin and Primitive States of the World, is to be allow'd for True, where 'tis fully agreeable to Scripture, Reason, and Philosophy." Newton, on the other hand, is less convinced than Whiston that the Bible must be the criterion by which one measures other ancient documents and, as Westfall has shown, in his manuscript "Theologiae Gentilis Origines Philosophicae" Newton places Egyptian records on an equal footing with the Bible. This attitude toward Scripture ultimately leads Newton to revise standard chronology, an action for which he is attacked by Whiston with great success. Because of Newton's attitude toward Scripture, Westfall has read Newton as a kind of deist.[18] On this one point – that is, Newton's equating the Bible with other ancient records – Westfall is correct. Nevertheless, Newton agrees that when properly interpreted the Bible accurately reveals both general creative divine providence and a specially provident God still directly active in creation and revealed through accounts in Scripture of fulfilled historical prophecies and accomplished miracles.

By examining Whiston's controversial works in the context of the historical circumstances of their origin, a much less paradoxical and more interesting figure emerges. A much firmer "comprehension" (*Erkennen*) and not merely an "understanding" is achieved, as Whiston, the man who lived through one of England's most turbulent periods (1667 to 1752), who was a renowned academic and Newtonian disciple before 1710 and often a laughingstock after that date, and who made a sustained attempt in all his writings to achieve a synthesis of Newtonian science, natural religion, revealed religion, Whiggish politics, Arian theology, and radical millennialism, marches forth from the bewildering array of his works to greet us. Without such an attempt to understand Whiston's controversies with orthodox Anglicans, coffeehouse deists, and even his former mentor, the great Newton, we are condemned to a truncated understanding of Newtonianism.

I

THE TEMPER AND TIMES OF A NEWTONIAN CONTROVERSIALIST

IN Section I of this chapter, I trace the historical details of Whiston's life, especially his opinions and attitudes about the controversies in which he took part and the people he knew. Only by first tracing this historical context is a fuller understanding of Whiston's intellectual achievements and of his role in shaping a genuinely Newtonian rapprochement between science and religion possible. In Section II, I sketch Whiston's impact in his own day and in ours.

I. The Primrose path

William Whiston was born in 1667 in Leicestershire during the reign of Charles II. He died in 1752 during the reign of George II at the age of eighty-four. During his lifetime he witnessed and often played a significant part in major events, including the party strife that followed the Glorious Revolution; the succession of Queen Anne and then of the house of Hanover; the establishment of the Board of Longitude; the rise of Newtonian science and its wider flowering as a religious and sociopolitical doctrine; the growth of the American colonies; and the rise of critical deism, which culminated in the religious skepticism of David Hume, who was forty-one when Whiston died. As if his own times were not sufficiently interesting, Whiston hoped and fully expected to see within his lifetime the restoration of the Jews, the Second Coming of Jesus, and the establishment on earth of the millennial paradise. Whiston's many works and, especially, his *Memoirs,* reveal one of the most fascinating figures of the eighteenth century, who is of crucial importance for an understanding of the Newtonian rapprochement between science and religion.[1]

Whiston's father, Josiah, had been a Presbyterian, but after the Restoration he conformed and retained his post as rector of Norton juxta Twycrosse. His father had been told by the apparitor that he must subscribe to the terms of the Restoration settlement or else "his Name must be put into the Roll of Refusers, or into his black Book, to be seen by those in Authority. The Consequence of which my Father so dreaded, that he did at last subscribe; but deeply repented it all the Days of his Life, and upon his Death-bed also."[2] The elder Whiston had no desire to "intermeddle" in affairs of state.

William Whiston apparently remembered his father's example and never chose preferment over the dictates of conscience. Whiston, his father's amanuensis, was "bred a scholar" from an early age and intended for a clergyman. Whether due to his long hours copying sermons and helping his blind father learn them or to his own "original *Stamina vitae*," Whiston observes that "I have been greatly subject to the *Flautus Hypochondriaci* in various Shapes all my Life long, although old Age, Temperance, Abstinence, and very great Exercise, have made it a great deal easier to me now for many Years."[3]

His early training enabled him to leave Tamworth School in less than two years and to enter Cambridge University in 1686, where even as an undergraduate he specialized in the study of mathematics, which he studied eight hours a day. In the aftermath of the Exclusion Crisis, Whiston records the rampant fear of "Popery and Persecution" among English Protestants that prevailed until "the Prince of *Orange* came to our Deliverance."[4]

Whiston received his B.A. in 1690 and in 1691 was elected to a fellowship at Clare Hall, Cambridge. In 1693, Whiston became a Master of Arts and contemplated taking holy orders as his father had intended. He determined not to receive his ordination at the hands of any bishop who was not satisfied in his conscience with the oaths to the new king and queen or from any bishop "how excellent soever, who had come into the Place of any who were not satisfied with the Oaths . . . and so had been deprived for preferring Conscience to Preferment."[5] Whiston's memory of his father's regret for subscribing to the Restoration articles of settlement was strong, and he renewed his own resolve to choose the Primrose path of conscience over worldly reputation or gain.

Fortunately, Whiston "pitch'd upon" Bishop William Lloyd, of Coventry and Litchfield, who, before the Glorious Revolution, had been bishop of St. Asaph and had been active in bringing William and Mary to the throne. Lloyd consented to ordain Whiston[6] and did so on December 28, 1693. Lloyd proved a major influence on Whiston's career as a Newtonian exegete of biblical prophecies. Whiston writes of Lloyd that

> this truly great and good Bishop . . . took me into his Bosom, and loved me, as I did him most sincerely; he understood the sacred Chronology, the Holy Scriptures, and particularly the Prophecies therein contain'd,

far better, I believe, than any *Jew* or *Christian* in the World before him; and whom I have heard *thank God for being able to read the Prophecies as he read History.*[7]

Following his ordination, Whiston returned to Cambridge to continue his studies in mathematics and also in the then popular Cartesian philosophy. Whiston adds,

> But it was not long before I, with immense pains, but no assistance, set myself, with the utmost zeal, to the study of Sir *Isaac Newton's* wonderful discoveries in his *Philosophiae Naturalis Principia Mathematica,* one or two of which lectures I had heard him read in the publick schools, though I understood them not at all at the time.[8]

In 1717, in his *Astronomical Principles of Religion, Natural and Reveal'd,* Whiston records the impact of Newton upon his thoughts:

> When in my younger days I had with great Difficulty and Pains attained to the Knowledge of the true system of the World, and of Sir Isaac Newton's wonderful Discoveries thereto relating, I was not only convinc'd, but deeply and surprizingly affected with the Consequences.

From his Newtonian enlightenment Whiston dated his "warm and zealous endeavours . . . for the Restoration of true Religion."[9] Whiston, in perhaps the most striking text of his *Memoirs,* traces the effect of his understanding of Newton's discovery of the law of gravity upon his own thought. He looks upon this "noble discovery"

> in an higher light than others, and as an eminent prelude and preparation to those happy *times of the restitution of all things, which God has spoken of by the mouth of all his holy prophets, since the world began,* Acts iii. 21 . . . . Nor can I forbear to wish, that my own most important discoveries concerning true religion, and primitive christianity, may succeed in the *second* place to his surprizing discoveries; and may together have such a divine blessing upon them, that the *Kingdoms of this world* . . . may soon *become the Kingdoms of our Lord.*[10]

Whiston also set himself up at this time as a tutor, with the encouragement of Archbishop Tillotson, who sent his nephew to be one of Whiston's eleven pupils. "Ill health" would not permit Whiston to pursue this employment, and he arranged for his good friend Richard Laughton, then chaplain to John Moore, the bishop of Norwich, to assume his tutoring duties. Bishop Moore then invited Whiston to become his chaplain in Laughton's place, in which capacity Whiston functioned from 1694 to 1698, while also still a Fellow at Clare Hall.

During this period Whiston wrote his first work, *A New Theory of the Earth, from its Original to the Consummation of all Things.* Shown in manuscript to Richard Bentley and Christopher Wren, "but chiefly laid before Sir *Isaac Newton* himself, on whose Principles it depended, and who well approved of it,"[11] the book was finally published in 1696. Fifteen hundred copies of this work were immediately printed, and it went through six editions, the last of which was published in 1755. Whiston's *New Theory* constitutes an attack on the deism, pessimism, and Cartesian rationalism expressed in Thomas Burnet's *Sacred Theory of the Earth,* which first appeared in Latin in two parts, the first of which was published in 1681 and the second in 1689. Whiston was especially eager to refute Burnet's allegorical interpretation of earth history as revealed in Scripture. It is in *A New Theory of the Earth* that Whiston shows how an orthodox Newtonian mathematician and scientist can be simultaneously a biblical scholar and exegete, and, as we shall see in Chapter 2, he developed these ideas further in his *Astronomical Principles of Religion, Natural and Reveal'd,* published in 1717.

Whiston's extreme moral rectitude and determination to follow his conscience occasionally reveals itself as a certain priggishness. For example, as chaplain to Bishop Moore it was Whiston's duty to present candidates for ordination "whose character was unexceptionable." He refused to present Bishop Stillingfleet's son as a candidate because he had heard "a very bad character as to his morals," but his delicate conscience was eased when his colleague Archdeacon Jeffries volunteered to present the young Stillingfleet because Jeffries had had "a better character of him than I had."[12] In another example, Whiston records an incident when, after being newly elected a Fellow in Clare Hall, he refused to support the candidacy of a former close friend who

> . . . thought at first that of the Electors the major Part were on the side of the Drinkers; and accordingly forsook his Sobriety, and For a Month or six Weeks drank hard with them at the Tavern, till we that were his old sober Friends saw it, and discarded him. . . . He at last found his Mistake, and that the sober Party were likely to be the Majority . . . and tried earnestly to recover his old Friends Votes, but to no Purpose.[13]

In 1698, Bishop Moore awarded Whiston the living of *"Lowestoft cum Kessingland* by the Seaside in *Suffolk"* (Samuel Clarke then became Moore's chaplain). At this vicarage in Suffolk Whiston had spiritual charge of two thousand souls. He was puritanically zealous in the performance of his pastoral duties in this small fishing village. He said early morning prayers, preached two sermons a day, and, in the summer, gave "catechetick lectures" in the evening to which "the *dissenters* would come, and by which I always thought I did more good than by my Sermons."[14] Whiston no doubt preferred Dissenters to his drunken ex-friend because of their adherence to their own pious convictions and principles, a trait of the Dissenters that he had always admired even before he became one.[15] Predictably, Whiston refused to help license a new alehouse in Lowestoft.[16]

Whiston resigned his living in 1701 when Isaac Newton invited him to return to Cambridge as his substitute. In 1696, Newton, Lucasian Professor of Mathematics at Cambridge, had been appointed Warden of the Mint, which was then located in the Tower of London. Newton accepted and moved to London in the spring of 1696, though he retained his chair at Cambridge. Early in 1701 Newton invited Whiston to lecture as his substitute with "the full profits of his place." When Newton finally relinquished the Lucasian Chair later that year, he made sure that Whiston would eventually be his successor by recommending him for that post to the heads of the colleges in Cambridge.[17]

From the time of his summons from Lowestoft to Cambridge in 1701 until his banishment from the university in 1710, Whiston pursued a fruitful academic career that combined his interest in biblical exegesis and his interest in Newtonian philosophy and mathematics. As Lucasian Professor, he gave the public university lectures on astronomy and mathematics. In 1702, Whiston published his *Short View of the Chronology of the Old Testament, and of the Harmony of the Four Evangelists.* In 1703, Whiston says, "I published my *third* book, which was *Tacquet's Euclid,* with *Selected Theorems of Archimedes,* and with the addition of *Practical Corollaries,* in Latin; for the use of young students in university."[18] In 1706, Whiston published *An Essay on the Revelation of Saint John, So far as concerns the Past and Present Times.* In 1707, he delivered the Boyle Lectures, which were published in 1708 as *The Accomplishment of Scripture Prophecy.* These lectures form the basis of one of Whiston's most important controversies, the debate with the deist Anthony Collins over the proper method of interpreting prophecy. Also in 1707 Whiston obtained Newton's permission to publish Newton's Lucasian Lectures on algebra[19] and first published his own *Praelectiones Astronomicae,* which he read in his public lectures at Cambridge.[20]

In addition to his wide-ranging publishing and lecturing activities, Whiston was active in the affairs of the university. He was instrumental in securing the newly endowed Plumian Professorship in astronomy and experimental philosophy for his friend and fellow Newtonian Roger Cotes, the editor of the famous second edition of Newton's *Principia,* published in 1713. Whiston records that "I was the only professor of mathematicks directly concerned in the choice, so my determination naturally had its weight among the rest of the electors." He modestly adds his reason for supporting Cotes: "I said, that I pretended myself to be not much inferior in mathematicks to the other candidates master, Dr. *Harris;* but confessed that I was but a child to Mr. *Cotes:* so the votes were unanimous for him."[21]

After Cotes came to Cambridge, Whiston gave courses of lectures with him in "hydrostaticks" and "pneumaticks" that featured the public performance of "philosophical experiments." These lectures later became the nucleus for some of the public lectures that Whiston gave "for many years" in London with Francis Hauksbee.[22] William Stukeley records that both he and Stephen Hales attended Whiston's Cambridge lectures on hydrostatics and pneumatics (see Chapter 5).[23]

He was particularly close to Samuel Clarke, his successor as chaplain to Bishop Moore. In 1705, Clarke published his own Boyle Lectures for the years 1704 and 1705 under the titles *Discourses Concerning the Being and Attributes of God; The Obligations of Natural Religion,* and *The Truth and Certainty of the Christian Revelation.* Whiston records an incident relating to Clarke's books that reveals Whiston's empirical turn of mind. When Clarke brought him a copy of the first volume of lectures, Whiston writes,

> I was in my Garden over against St. *Peter's* College in *Cambridge,* where I then lived. Now I perceiv'd that in these Sermons he had dealt a great deal in Abstract and metaphysick Reasonings. I therefore asked him how he ventur'd into such Subtilties, which I never durst meddle with? And showing him a Nettle, or the like contemptible Weed in my Garden, I told him, "that Weed contained better Arguments for the Being and Attributes of God than all his Metaphysicks." Mr. *Clarke* confess'd it to be so: but alleg'd for himself, "that since such Philosophers as *Hobbs* and *Spinoza* had made use of those kind of subtilties *against;* he thought proper to show that the like way of Reasoning might be made better use of *on the Side* of Religion." Which Reason or Excuse I allow'd not to be inconsiderable . . . My own Opinion is, that perhaps *Angels* or some of the Orders of rational Beings superior to them, may be able to reason a great way *a Priori* . . . and from Metaphysicks [but] I do not perceive that we *Men,* in our present imperfect state, can do so . . . *Quae supra nos nihil ad nos.*[24]

Another discussion with Clarke about this time put in train a sequence of events that transformed Whiston's life. Around 1705, Whiston learned that Clarke doubted the validity of the Athanasian doctrine of the Trinity and never read the Athanasian creed in his parish near Norwich. Whiston records that after a conference with Newton at about the same time, he "returned much more inclin'd to what has been of late called Arianism."[25] Following his 1707 Boyle Lectures, Whiston had time to pursue research on this topic. Writing in the year 1711, Whiston states that "about the

Month of *February,* 1708, I was desir'd by a Friend or two to draw up such a Method, or Directions for the Study of the Divinity, as I us'd in Conversation to them and others, as the only way for the Union of Christians and Restauration of the Primitive Faith."[26]

The result of this research supported the antitrinitarian school of Clarke and Newton. A study of the earliest Christian records of the first four centuries convinced Whiston that the trinitarian doctrine of the complete identity of the Son with the Father was a false accretion to original, "primitive" Christianity. The original doctrine of the Christian church and, indeed, of Christ himself, was that the Son was not a coexisting, consubstantial being that existed prior to the Incarnation from eternity but possessed rather "a *Metaphysick existence, in potentia,* or in the like higher and sublimer Manner *in* the Father as his *Wisdom* or *Word* before his real *Creation* or *Generation.*"[27] The corollary to this discovery about the true nature of Christ was the discovery of how the church made into dogma the false trinitarian doctrine. The introduction of trinitarianism into Christianity dated from the latter part of the second century A.D., "after Philosophy was come into the Church." But it only "advanc'd to a mighty system in the fourth, under the Conduct of *Athanasius.*"[28] Whiston's perusal of the *Apostolical Constitutions* in the summer of 1708 confirmed his Arian hypothesis. This work was purported to be a collection of ecclesiastical regulations drawn up by the Apostles. He was introduced to the *Constitutions* by his friend Richard Allin, Fellow of Sidney College, Cambridge. Whiston, along with the rest of the learned world, had formerly regarded it as a forgery. Now, after close examination with Allin, he felt that it truly contained, as its title proclaimed, the original governing laws and doctrines of the church, personally delivered by Christ to the Apostles. It was, incidentally, in this document that Whiston discovered evidence for his view that priests ought not to marry more than once, the doctrine that caused the Rev. Dr. Primrose so much strife.

With his Arian hypothesis confirmed by historical research in the earliest Christian documents, Whiston, in July of 1708, wrote to inform the archbishops of York and Canterbury that as a result of a fourteen-hundred-year conspiracy, the church had been teaching false doctrine and that he, William Whiston, could prove its falsity and also show how to reform Christian teaching by bringing it into conformity with the original. For one so often charged with having a "warm head," Whiston was remarkably circumspect in seeking the advice of the two archbishops. He clearly recognized that he might be wrong and stated that he had no wish to upset the "Unity and Quiet" of the church of which he was a "peaceable Member." Moreover, he insisted that

I would have this Matter calmly and fairly debated and settled by the Learned, before it comes into the Hands of the Ignorant; Upon all these Accounts, I humbly propose it to your Graces Consideration, what Way I should take in the particular Management of this Matter? My own Thoughts are . . . to have some Copies transcrib'd, or rather a few Printed, for the use of the Learned.[29]

Archbishop Tenison of Canterbury replied on July 24, urging Whiston to be cautious and to think twice before proceeding but in no way prohibiting his scheme.[30] Archbishop Sharp of York urged Whiston, in a letter dated August 6, 1708, to delay his project

at least so long as till you have had opportunity of talking freely about this matter with your friends at *London;* which you may have in the Parliament-time, if you will then be so kind as to make a journey thither. A great many things may be offer'd in Discourse, for the Conviction of either of the differing Parties, which cannot be so easily writ in Letters.[31]

Whiston's former mentor, Bishop Lloyd of Worcester, also wrote to dissuade Whiston from engaging any further with the "Socinian" heresy and warned that if he continued on this course it would be necessary for Lloyd to "break friendship" with him "once for all . . . God forbid it should ever come to this!"[32] In August 1708, Whiston applied to the vice-chancellor at Cambridge for a license to print at Cambridge the results of his research, his "small, imperfect" *Essay upon the Apostolical Constitutions.* Dr. Laney sent messengers refusing such a license, because he did not think the work "orthodox."[33]

Whiston finally published some of the results of his research into primitive Christianity for the first time in 1709 in *Sermons and Essays Upon Several Subjects.* The tenth sermon, "Advice for the Study of Divinity: With Directions for the Choice of a Small Theological Library," contains his methodology. His argument is that since the Reformation and the rejection of papal infallibility, Protestants have claimed to go back to the original sources of revelation, but in many cases of dogma they stop with later documents rather than going all the way back to the earliest, purest records that are nearest the divine source. This methodological principle underlies Whiston's *New Theory of the Earth.* In the eleventh and last essay, Whiston published the explosive results of this method when applied to the doctrine of the Trinity. True to his letter to the archbishops, he sought to keep the matter in the calm and fair hands of the learned, and so wrote the essay in Latin. It is entitled "Incert: Auctoris De Regula Veritatis, sive Fidei: vulgo, Novatiani De Trinitate Liver."[34] When Whiston's friends Dr. Richard Laughton and a Mr. Priest heard in the summer of

1709 what Whiston was "going about," they came to his house to stop him from going any further, arguing the "hazards and dangers" for Whiston and his family in such a course. Whiston replied, "I have studied these points to the bottom, and am thoroughly satisfied the christian church has been long and grossly cheated in them; and, by God's blessing, if it be in my power, it shall be cheated no longer."[35] He made a similar response to a plea from Bentley.[36] Whiston also at this time denied the coeternity and consubstantiality of Christ in the public catechetical lectures that he delivered in the parish church of St. Clement's in Cambridge.[37]

Finally, in 1710, Whiston was summoned before the heads of the houses at Cambridge and charged with breaking the forty-fifth statute of the university, which prohibited the teaching of doctrines at the university contrary to established Anglican belief.

Although he had endeavored to keep his doctrine in Latin, his accusers found sufficient evidence in the English sermons and essays to convince them that Whiston had taught that "the Father alone is the One God of the Christian Religion, in opposition to the Three Divine Persons, Father, Son, and Holy Ghost, being the One God of the Christian Religion."[38]

Found guilty, Whiston was deprived of his professorship and banished from the university. At the age of forty-three, with a wife and four children to support, he moved to London. His legal troubles on the charge of heresy were just beginning. As Archbishop Sharp had tried to warn him in his letter of August 6, 1708, the political climate was most unfavorable for Whiston's vocal heterodoxy, despite Whiston's many friends during "Parliament-time." A tide of High Church Tory feeling was beginning to swell in 1708 with the trial of Dr. Henry Sacheverell for "high crimes and misdeameanours." His high-flying Tory supporters raised the rallying cry of "Church in Danger," and in 1710, when the Tory tide swept out the Godolphin ministry together with the Whig junto and swept in the Tory ministry of Harley and Bolingbroke, pressure mounted to stage a Tory counterrevolution. High-flying Anglican tories led by Francis Atterbury sought to reestablish High Church Anglican judicial authority in church and state through the Convocation of 1710–11. Whiston was charged with heresy before this body, but when Convocation broke up in 1712 it had failed to convict him. The frustrated High Church rector of St. Anne's, Westminster, John Pelling, who had pushed for conviction, next initiated heresy proceedings before the Court of Delegates. This attempt to secure conviction failed too. Whiston's defense in these London heresy trials, his Whig supporters, and his developing Whig political theory will be explored in detail in Chapter 4. Despite all these legal proceedings against him in Cambridge and in London, which lasted nearly five years, Whiston notes in his *Memoirs* that he does not remember losing more than two or three nights' sleep on their account.[39] Of his prolonged conflict with the established church, Whiston observes that his Whig friend "Sir *Richard Steel* hit the mark, when he thus distinguished the two principal churches in *Christendom,* the church of *Rome* and the Church of *England;* that the former pretended to be infallible; and the latter to be *always in the right.*"[40]

Whiston's banishment from Cambridge formed the major turning point in his life. Up to 1710, he was controversial but respectable. His vocal support for the Arian cause tainted his reputation and deprived him of his academic position, as he realized it would. After 1710, because of his act of conscience, he was forced to move down to London and scramble madly for a living. To support his family, he pursued a variety of lecturing and publishing schemes while adhering steadfastly to the dictates of his conscience. He recounts the following anecdote, which clearly reveals his temperament:

> When I was once talking with the lord chief justice *King,* one brought up among the dissenters at *Exeter,* under a most religious, christian, and learned education, we fell into a debate about signing articles, which we did not believe for preferment; which he openly justified, and pleaded for it, that *We must not lose our usefulness for scruples.* [Strange doctrine in the mouth of one bred up among dissenters! Whose whole dissent from the legally established church was built on scruples.] I reply'd, that I was sorry to hear his lordship say so; and desired to know, whether, in their courts, they allowed of such prevarication or not? He answered, They did not allow of it. Which produced this rejoinder from me, "Suppose God Almighty should be as just in the next world, as my lord chief justice is in this, where are we then?" To which he made no answer. And to which the late Queen *Caroline* added, when I told her the story, Mr. Whiston, *no answer was to be made to it.*[41]

In London, Whiston embarked upon a publishing career even more prolific than that of the period of his Cambridge professorship. His Arianism now strident, he engaged in extensive polemics on behalf of primitive Christianity. In 1711 he published, in four volumes, his *Primitive Christianity Reviv'd,* his full-scale defense of the *Apostolical Constitutions* and of the Arian doctrine. Whiston was repeatedly attacked, and he always counter-attacked.[42] In 1715 Whiston founded a Society for Promoting Primitive Christianity, which met weekly at his house in Hatton Garden until 1717. Whiston refers to his house as the "Primitive Library." Meetings consisted of ten or twelve "Christians of all Persuasions" and included Dr. John Gale, Thomas Chubb, Benjamin Hoadley, Thomas Emlyn, and Arthur Onslow, later a Speaker of the House.[43] Also present was Thomas

Rundle, who was "very ready to join with me for restoring primitive Christianity." Much to Whiston's disgust, Rundle (and Hoadley) ultimately signed the Thirty-nine Articles for the sake of preferment. Rundle then argued that the *Apostolical Consitutions* dated from the fourth century, rather than from the time of the Apostles. Whiston retorts, "Make but Dr. *Rundle* dean of Durham, and they will not be written 'till the fifth century."[44] There is one report saying that it is "probable" that Queen Caroline sometimes attended these gatherings incognito.[45] Queen Caroline was introduced to Whiston by Samuel Clarke, who came to her weekly to discuss theology and philosophy and occasionally brought Whiston with him. The queen enjoyed Whiston's plainspokenness and accepted his reproofs in good humor.[46] She gave him a stipend of forty pounds "clear" a year, which was continued by George II after her death. This annuity, with the twenty pounds a year provided him for life by Sir Joseph Jekyll, a leading Whig at Court and Master of the Rolls, and the money he made "with eclipses, comets, [and] lectures of several sorts in *London* and elsewhere" enabled him to "go on all a long comfortably" with his studies.[47]

Whiston's public lectures were a source not only of income but also of influence and yet more notoriety. One of Whiston's most influential course of lectures was probably his first. These lectures were given beginning in January 1714 and had been arranged by Joseph Addison and, especially, Richard Steele. Steele had been introduced to Whiston by Henry Newman, secretary of the Society for Promoting Christian Knowledge. Whiston was a long-time member of the Society but had quit to prevent stigmatizing that "most worthy" organization with his heretical stamp and thus driving away contributors.[48] In his letter recommending Whiston to Steele, Newman adds

> I only beg leave to suggest one thing to you when he does [lecture] because it will come with more authority from you than perhaps any man in the kingdom beside, and that is that you will be pleas'd to conjure him Silence upon all Topicks foreign to the Mathematick in his Lectures at the Coffee house. He has an Itch to be venting his Notions about Baptism & the Arian Doctrine but your authority can restrain him at least while he is under your Guardianship.[49]

In his *Memoirs* Whiston credits both Steele and Addison, his "particular friend,"[50] with setting up "many astronomical lectures at Mr. *Button's* coffeehouse, near Covent Garden to the agreeable entertainment of a good number of curious persons."[51] Such scientific lectures to paying audiences were enormously important in the popularization of science at first in London and, after midcentury, in the provinces.[52] Whiston's "principal auditor" at the lecture series that began in January 1714 was the extremely important Whig politician and war hero Lord James Stanhope.[53] On April 22, 1715, there occurred a total eclipse of the sun that proved a windfall for Whiston's finances:

> I myself by my lectures before; by the sale of my schemes before and after; by the generous presents of my numerous and noble audience; who at the recommendation of my great friend, the lord *Stanhope,* then secretary of state, gave me a guinea apiece; by the very uncommon present of my great benefactors, the duke of NewCastle; and of five guineas at night from the lord *Godolphin;* gained in all about 120 l. by it. Which, in the circumstances I then was, and have since been, destitute of all preferment was a very seasonable and plentiful supply: and, as I reckoned, maintained me and my family for a whole year together.[54]

Whiston continued giving public lectures and "courses of experiment" throughout his life. Astronomy, especially eclipses, continued to be one of Whiston's main topics. Lectures on biblical chronology and the imminent millennium became important as well.[55] In 1726, Whiston designed and had constructed scale models of the tabernacle of Moses and the Temple of Ezekiel at Jerusalem. At London, Bristol, Bath, and Tunbridge Wells he used these models as the basis for millennial lectures about the recall of the Jews, the rebuilding of the Temple, and the Second Coming. In 1746, at age seventy-nine, he repeated these lectures at Hackney and Tunbridge Wells, to the "great satisfaction" of his audiences. He is pictured in a print from the *Richardson Correspondence* in 1748 on the main street of Tunbridge Wells with the other "remarkable characters" who frequented the resort at that time. With Whiston on the street are Arthur Onslow, his old colleague from the Society for Promoting Primitive Christianity and now Speaker of the House; Samuel Richardson; and Dr. Samuel Johnson. Whiston's brisk and vigorous attitude reveals that walking was no doubt part of the "vigorous exercise" that, along with temperance, abstinence, and old age, had helped him combat his condition of "Flautus Hypochondriaci[56] (see Figure 2).

In order to earn money, Whiston not only gave publicly subscribed lectures on experimental science, astronomy, mathematics, and the millennium; did private tutoring;[57] and published such important works as *Astronomical Principles of Religion, Natural and Reveal'd* (1717); he also invented at this time a way to discover longitude for ships at sea. In collaboration with Humphrey Ditton, a mathematics teacher at Christ's Hospital, Whiston attempted to generate interest in Parliament for the reward of a prize for such a discovery. A letter by Whiston and Ditton describing the project was published by Addison in the *Guardian* (no. 107, July 11, 1713), prefaced with a highly laudatory introduction by Addison.[58] In April 1714, Whiston and Ditton petitioned Parliament to establish such a prize. In May 1714, a group of merchant and Navy mariners and "merchants of London" put forward a similar petition. A committee of the House summoned such renowned scientists as Newton, Halley, Cotes, and Samuel Clarke to advise them whether such a prize ought to be offered. These learned men reported their assessment of Whiston's and Ditton's method to the House committee on June 11, 1714. Cotes, Clarke, and Halley all spoke briefly, and no record of their opinions is preserved in the House records.[59] Newton spoke from a prepared text that summarized the four basic methods for determining longitude at sea: (1) calculating time with a perfect timepiece; (2) observing the eclipses of the satellites of Jupiter; (3) observing the positions of the moon; and (4) the project of Whiston and Ditton announced in Addison's *Guardian* a year before, the details of which had been privately communicated to the House committee and its scientific confreres.

Whiston's description of the June 11, 1714, meeting, which he and Ditton attended, is fascinating. The committee's purpose was only to determine whether a prize ought to be offered. After the verbal reports of Cotes, Halley, and Clarke, and Newton's scripted reading of the four methods and their disadvantages, the chairman of the House committee, who was against a cash prize, told Newton that unless he declared in favor of the practicality of Whiston's and Ditton's scheme, as Cotes and Halley had apparently done, no prize would be established. Newton sat silent. Whiston then spoke out, urging Newton to endorse the practicality of their scheme, at least near the shore. Newton then did speak, repeating what Whiston had said. The House committee, which included Whiston's friends Lord James Stanhope and Joseph Addison, then voted unanimously to fund a prize. Addison and Stanhope and some other members to the committee then drafted a bill "for Providing a Publick Reward for such Person or Persons as shall Discover the Longitude at Sea," which passed the Commons on July 3, 1714, the Lords on July 8, and was signed into law by Queen Anne on July 20,[60] twelve days prior to her death and a few weeks after Whiston had made known in public the details of his and Ditton's scheme.[61] The commissioners who were appointed under the terms of this act came to be known as the Board of Longitude.

Whiston's and Ditton's project, made public in the summer of 1714, involved anchoring station ships along the trade routes. Each midnight (by Peak of Tenerife time) each ship would fire a star shell set to burst at a fixed altitude. A compass would provide the bearing to the station ship. Distance from the station ship could be obtained either by noting the length of time it took the sound to reach the observer or by observing the angle subtended by the bursting star shell, which was set to explode at 6,440 feet. Whiston did not win the prize with this or with his other two methods, one involving the inclination of dipping needles (1721),[62] and one based on observation of eclipses of the moons of Jupiter (1738). But he had played a key role in lobbying effectively for the establishment of the Board of Longitude, which eventually awarded the full prize of twenty thousand pounds to John Harrison, between 1737 and 1773, for his continual refinements during that period of a practical and accurate marine chronometer.[63] Harrison specifically set out to perfect a marine chronometer in order to win the huge prize.

As we have seen, Newton supported Whiston's testimony before the House committee, but perhaps with some hesitation. Earlier, his support of Whiston's career had been generous. But, whether because of Whiston's enthusiastic and strident militancy in behalf of Arianism and Newton's own natively cautious temperament that made him fearful for his social and political position in London, or because of the allegation that Whiston's nephew (whom Newton employed) had stolen a large sum of money from Newton's house, Newton ultimately ostracized Whiston and excluded him from his circle of anointed disciples.[64]

Whiston's own account is, as usual, as reliable as any. While they were having coffee with some friends at Child's Coffee House in St. Paul's Churchyard in 1720, Edmond Halley asked Whiston why Whiston was not a member of the Royal Society. Whiston replied

> ... because they durst not choose an *Heretick*. Upon which Dr. Halley said to Sir *Hans Sloane,* that if he would propose me, he would second it: which was done accordingly. . . . When Sir Isaac Newton, the president heard this, he was greatly concern'd; and, by what I then learn'd, closeted some of the members, in order to get clear of me; and told them, that if I was chosen a member, he would not be president. Whereupon, by a pretence of deficiency in the form of proceeding, the proposal was dropped. . . . if the reader desires to know the reason of Sir Isaac Newton's unwillingness to have me a member, he must take notice, that as his making me first his deputy, and giving me the full profits of the place, brought me to be a candidate, as his recommendation of me to the heads of colleges in *Cambridge,* made me his successor; so did I enjoy a large portion of his favour for twenty years together. But he then perceiving that I could not do as his other darling friends did, that is, learn of him, without contradicting him, he could not, in his old age, bear such con-

tradiction; and so he was afraid of me the last thirteen years of his life. . . . He was of the most fearful, cautious, and suspicious temper, that I ever knew.[65]

The Royal Society never refused to allow him to present before them his "papers or instruments," as he did with his method for determining the longitude by the dipping needle in 1721. Though from quite early in their relationship, Whiston had argued with Newton in private about how to interpret particular biblical prophecies,[66] Whiston's independent turn of mind finally became more than Newton could bear. Yet even after Newton cast Whiston off, Whiston refused to enter into public debate with him until after Newton's death. Whiston states that even if he had completed what he regarded as an utterly devastating critique of Newton's revision of historical (biblical) chronology while Newton was still alive, he would have suppressed it, "because I knew his temper so well, that I should have expected it would have killed him." Whiston declares emphatically that "I would have done nothing that might bring that great man's *grey hairs with sorrow to the grave*."[67] Whiston's ego was clearly as powerful as Newton's.

Whiston remained active as a writer and lecturer until the end of his life. In 1737 he published one of his most successful works, his translation of the works of Josephus, *The Genuine Works of Flavius Josephus the Jewish Historian,* which has been continually in print for more than two centuries and was reprinted as recently as 1981. In his last years he planned to work specifically on the circumstantial details of the events preceding the coming of the millennium. Following the description that he gives in his *Memoirs* of his 1746 lectures at Hackney and Tunbridge Wells on Ezekiel's Temple, Whiston says,

> I intend many more the like lectures, and those that are a preparatory to the restoration of the *Jews* hereafter, while it shall please God to continue my health and abilities to go through them. And this I take to be my peculiar business at present; Since I have, I think, plainly discovered, that it will not be many years before the Messiah will come for the restoration of the *Jews,* and the first resurrection, when the last of these temples, the temple of *Ezekiel,* will be built upon *Mount Sion.*[68]

As we shall see in Chapter 5, in 1750, at the age of eighty-two, Whiston continued his millennial lecturing to crowds in London upon the occasion of the minor earth tremors there on February 8 and again, one month later, on March 8. But by this time, this particular aspect of Whiston's thought had become laughable to many.

On August 22, 1752, the pious controversialist died after a week's illness at the age of eighty-four years, eight months, and thirteen days. Fourteen years later, in 1766, Goldsmith published his tale of the Rev. Dr. Primrose. And four years earlier, in 1762, Hume had published his essays, "Of a Particular Providence and a Future State" and "Of Miracles."

II. Whiston's impact and reputation

Whiston lived a very long time; his literary reputation and his impact on the thought of his times changed with the times. Early in his career, the popularity of his *New Theory of the Earth* caused Newton to bring him to Cambridge. Locke, writing to Molyneaux his evaluation of Whiston's book, gave the opinion of the learned world in 1696:

> You desire to know what the opinion of the ingenious is concerning Mr. Whiston's book. I have not heard any one of my acquaintance speak of it, but with great commendations, as I think it deserves, and truly I think he is more to be admired that he has laid down an hypothesis whereby he has explained so many wonderful, and before inexplicable things in the great changes of this globe, than that some of them should not easily go down with some men; when the whole was entirely new to all. He is one of those sort of writers that I always fancy should be most encouraged; I am always for the builders, who bring some addition to our knowledge, or, at least, some new thing to our thoughts.

In 1703, Locke wrote a letter entitled "The best Method of studying and Interpreting the Scriptures" in which he highly recommended Whiston's *Harmony of the Four Evangelists* as the best means for achieving a thorough understanding of the Gospels.[69]

Following Whiston's banishment from Cambridge and the period of intense controversy that surrounded the legal proceedings against him in London for heresy, his reputation suffered, especially at the hands of the Scriblerians. Around 1708, Alexander Pope refers to the "wicked works of Whiston" in a verse letter to Cromwell.[70] A distasteful piece of scatological doggerel appeared after the publication of Whiston's and Ditton's stationship project for the determination of the longitude in 1714. These verses were by another Scriblerian, James Parnell, who wrote them to "jostle down" Whiston for his Arianism, as we learn in a letter from Sir Richard Cox, former lord chancellor of Ireland, to Edward Southwell, the English secretary of state. Cox writes, "Archdeacon Parnell has made the following dirty lines, which are valued, because they Ridicule the Confident Arrian Whiston."[71] The most extraordinary thing about his extraordinary set of verses is that they seemed to be appreciated because they belittled an Arian heretic, rather than a "projector." The Scottish Kirk burned *A New Theory of the Earth,* not because of any errors contained in that work but because its author was an Arian.[72] Jonathan Swift satirized all projects for determining the longitude in *Gulliver's Travels* and was critical of them in his correspondence.[73] John Arbuthnot, in his book about Woodward, referred to Whiston as a "late ingenious Writer" but complained in a letter to Swift that Whiston's published longitude scheme had "spoiled one of my papers of Scriblerus, which was a proposal for the longitude, not very unlike his."[74] Swift replies that "it was a malicious satire of yours upon Whiston, that what you intended as a ridicule, should be any way struck upon him for a reality."[75] John Gay also joined in by ridiculing Whiston's millennial lectures in "A True and Faithful Narrative of What pass'd in London," in which Whiston supposedly prophesies imminent cometary doom (see Chapter 5). The idea of a cometary apocalypse is also ridiculed in detail by Swift in *Gulliver's Travels,* where the ivory tower Laputans take the idea quite seriously and discuss it in great and learned detail.[76]

But Whiston did influence Pope in one very positive way. Nicolson and Rousseau have drawn welcome attention to the "great stimulus" of Whiston's conversations with Pope upon Pope's poetic imagination. In a letter to John Caryll dated August 14, 1713, Pope writes:

> You can't wonder my thoughts are scarce consistent, when I tell you how they are distracted. This minute, perhaps, I am above the stars, with a thousand systems round about me, looking forward into the vast abyss of eternity, and losing my whole comprehension in the boundless spaces of the extended Creation, in dialogues with W[histon] and the astronomers; the next moment I am below all trifles, even grovelling with T[idcombe] in the very center of nonsense: now am I recreating my mind with the brisk sallies and quick turns of wit, which Mr. Steele in his liveliest and freest humours darts about him.[77]

One of the most important effects of Whiston's intellectual career was his influence upon the burgeoning Unitarian movement in England in the early eighteenth century. The attempt by the High Church Tories to convict Whiston for heresy before the court of Convocation in 1710 and 1711 shows that they regarded his vocal publication of his Arian views as a real threat. In 1710, Whiston directly influenced Joseph Hallet, a schoolmaster who wrote to Whiston after reading the "Directions for the Study of Divinity" in the 1709 *Sermons and Essays.* Hallet requested "further direction" but begged Whiston not to make their correspondence public, so that Hallet might avoid the "ruin" that had befallen Whiston.[78] Whiston exercised a similar influence over James Peirce and Hopton Haynes, Newton's deputy at the Mint.[79] These men, in turn, directly influenced many Dissenters who, in the course of the eighteenth century, were moving away from the orthodox doctrine of the Trinity. Whiston also was instrumental in the career of Thomas Chubb. Whiston's *Primitive Christianity Reviv'd* (1711) prompted Chubb to produce his own *Eight Arguments on the Supremacy of the Father,* which he took to London and gave to Whiston. Whiston acted as editor when Chubb published this work in 1714 and introduced him to Sir Joseph Jekyll, who became his patron. Chubb also came to Whiston's "Primitive Library" in Hatton Garden, where, at meetings of the Society for Promoting Primitive Christianity, he met Thomas Emlyn, the first Unitarian minister in England; Arthur Onslow, later Speaker of the House; John Gale; and Benjamin Hoadley.[80] If the rumor is true that Princess Caroline visited these gatherings, possibly Chubb met her there too. Whiston influenced the people who were the architects of English Unitarianism indirectly, by standing up for Arianism in the face of orthodox persecution, and directly through his activities at the "Primitive Library." Throughout his life he advised antitrinitarians in England.[81]

Whiston was most widely respected as a scientist, and his ability to explain astronomy exercised influence, as we have seen in the case of Pope.[82] Whiston's linkage of astronomical phenomena with prophecy continued to exert some influence, even after the savage ministrations of Gay and Hogarth (see Figure 1). In 1779, the date of the publication of Hume's *Dialogues Concerning Natural Religion,* Dr. Johnson's friend, Mrs. Thrale, wrote of Whiston's effect on her imagination in enthusiastic lines reminiscent of Pope's reaction to Whiston in their "dialogues" about the great "abyss" sixty-five years earlier:

> There is no Reading that so changes the Scene upon one, and carries one so completely out of one's self I think, as Astronomical Speculation: unless indeed the Study of the Ancient prophecies and modern calculations of this World's final Dissolution: when we read Burnet on the Conflagration, or Whiston on the expected Comet how little seem the Common Objects of our Care! . . . When Heaven is half open'd to our Eyes by Prophecy, & this little Seat of Tumult seems to recede from our Sight; 'tis then that the Soul feels her elastic force . . . & expatiates in the Speculations for which She was originally formed.[83]

But by the time that Goldsmith made Whiston's primitive Christian doctrine that it was unlawful for a priest to take a second wife after the death of the first the focus of *The Vicar of Wakefield,* Whiston was considered, like Primrose, to be an eminently parodiable, even pathetic, religious quack.[84]

From the beginning of his career, Whiston and his works excited the curiosity of continental writers in both science and religion. In 1717, Michel de La Roche, the editor of *Bibliothèque angloise,* requested that Whiston write an abstract of the *New Theory*.[85] When the French Academician Nicolas Fréret wrote his attack on Newton's truncated biblical chronology in *Defense de la chronologie,* he happily relied upon Whiston's own "confutation" of Newton.[86] Still, Whiston was most famous for his *New Theory of the Earth,* which was lengthily described at midcentury by the Comte de Buffon.[87] Delambre summarized Whiston's theory in 1827 and recounted Whiston's influence on Maupertius and Lalande in the mid-eighteenth century.[88]

In Germany the *New Theory* received widespread prominence soon after its publication[89] and became even more famous following its German translation in 1713.[90] Buddeus commented on Whiston's theory in 1715.[91] Andrew Dickson White notes the impact and institutionalization of Whiston's earth theory in Germany by midcentury:

> The theories of Whiston and Burnet found wide acceptance also in Germany, mainly through the all powerful mediation of Gottsched, so long, from his professor's chair at Leipsic, the dictator of orthodox thought, who not only wrote a brief tractate of his own upon the subject, but furnished a voluminous historical introduction to the more elaborate treatise of Heyn. In this book, which appeared at Leipsic in 1742, the agency of comets in the creation, the flood, and the final destruction of the world is fully proved.[92]

Whiston's works were so well known in Germany by midcentury that his German fame occasioned his writing his *Memoirs.* In 1746 a "Hanoverian scholar" brought Whiston a biographical piece that he had written about Humphrey Ditton, Whiston's coauthor in the 1714 longitude book, and asked Whiston to examine and correct it. Whiston states that

> when this short account was shewed me in *English,* I perused it, and found no small parts of it to be false; and so I was forced to write it almost all new. . . . About a week afterward came another *Hanoverian* scholar to me, and desired me to write my own life: For he said, that also had been written in *Germany;* but as was now found, with several falsities likewise. My answer was, that though I had long ago put upon this, I had not hitherto inclined to do it; yet rather than go down to posterity with such falsities, perhaps I might set about it; as I did immediately.[93]

We are indebted to these Hanoverian scholars, then, for Whiston's fascinating, insightful, and diffuse personal history of his life and times.

George Berkeley included Whiston's *New Theory of the Earth* and *The Accomplishment of Scripture Prophecy* in the books he brought with him to the American colonies in 1728. With this library, Berkeley intended to establish a college in Bermuda for the training of clergymen and the education of Indians. When Berkeley abandoned this scheme in 1733, he donated these books to Yale.[94] In America, the *New Theory* met with its usual acclaim,[95] while both Whiston's chiliastic religious writings and his work on astronomy appealed strongly to the millennialist president of Yale, Ezra Stiles, who learned about them there in 1744.[96] In the mid-nineteenth century, Herman Melville, in his novel *Mardi,* was still able to utilize to good effect Whiston's striking vision of the earth after the final judgment, when the earth becomes once again a comet in a radically elliptical orbit on which the damned hurtle between the freezing depths of deep space and the boiling regions near the sun.[97]

Whiston's reputation today is that of an eccentric figure who, like Primrose, evokes a measure of amused pity. Leslie Stephen summed it up by describing Whiston as "a man of very acute but ill-balanced intellect."[98] The *Encyclopaedia Britannica* article from 1910–11 also adopts this view. It reads:

> Whiston is a striking example of the association of an entirely paradoxical bent of mind with proficiency in the exact sciences. He also illustrates the possibility of arriving at rationalistic conclusions in theology without the slightest tincture of the rationalistic temper.[99]

I do not wish to vindicate Whiston's intellectual career or to prove that he has been wronged in our modern interpretations. Indeed, they rest in part on the judgments of some of Whiston's contemporaries and so are not purely anachronistic. But such interpretations are only partially correct, because they first bifurcate Whiston into a scientist and a theologian and then proceed, on the basis of tendencies already fully developed at the end of Whiston's long life, to ridicule his theology and the idea that his theological position could have had any real influence, while simultaneously underemphasizing his contributions to science.

Whiston's latest biographer, while acknowledging that Whiston "maintained concurrently an active interest as much in scientific as in religious studies,"[100] has attempted to emphasize his "scientific" work in geology, his attempts to discover a method of fixing longitude, and his many other "scientific" books and lectures. One chapter is devoted to the Whiston who appears to this biographer as "a religious enthusiast who pursued ideas which appear very obscure to a twentieth century reader."[101]

Whiston's apparently obscure and often unsettling works cannot so easily be divided into the rationally sound and the irrationally unsound. Whiston is, at one and the same time, a "serious" and rational earth theorist, as well as the precursor of Immanuel Velikovsky.[102] He is simultaneously an arch-heretic and the man whose translation of Josephus is still in print and widely read in modern seminaries.[103]

Through an examination of a representative body of Whiston's controversial works in the historical conditions of their origins, a much less paradoxical figure emerges. Nevertheless, Whiston remains a man very much out of step with the world around him both in his day and in ours. Benjamin Hoadley, Whiston's friend and (if we are to believe Whiston) a dissimulating antitrinitarian, has left us one of the most sympathetic portraits of Whiston, one that well catches the characteristic temper of the man. In a long discussion addressed to the pope, explaining the lack of religious tolerance in England, Hoadley writes:

> In *England*, it is not all the other Excellencies in the World, united in one Man, that can guard him against the fatal Consequences of Heresy, or differing, in some Opinions, from the Current Notions of our World; especially if those Opinions are such as are allowed to be Mysterious and Inexplicable. We have now an Instance of One, or Two, Learned, and otherwise Good Men, who have thought it their Duty, (as they themselves say) to step aside out of the common Path. And what their Fate will be, Time must shew. At present, the Zeal (as it is called) of their Adversaries prevails. The Fire is kindled, and how far it will consume, or where it will stop, God only knows. But the Case of *One* of them (which will give Your Holiness some Notion how we stand affected) is very remarkable. For, not to mention his Good Life, (which is look'd upon but as a *Trifle,* common to almost all Modern Hereticks;) tho' his Religion is mix'd up with a good deal of *Kalendar* and *Rubrical* Piety; tho' He hath his Stated Fasts and Feasts, which He observes with the greatest Devotion, tho' He is Zealous for Building of Churches in the Apostolic Form of a Ship, with all Accomodations for Order and Decency; tho' he is for the Use of *Oyl,* and the *Trine Immersion* in *Baptism,* and for *Water* mix'd with *Wine,* in the *other Sacrament;* tho' he is very warm for Believing in Christ towards the *East,* and Renouncing the Devil towards the *West;* tho' He hath laid them a Foundation for *Independent Church-Power,* in the *Decrees* of the *Apostles* themselves; nay, tho' He joins with them in beating down Human Reason, when it would pretend to judge in Matters of Religion; and resigns to them all the Preferments in the Land, from *Dover* to *Berwick* upon *Tweed;* Yet all will not do: He holds the *Son* to be *Inferior* to the *Father,* and *Created* by *Him,* tho' a Being of more Glorious Perfections: And upon this Account, He must not enjoy, even the Poverty which he hath chose, in quiet. And if this be *His* Case, what hath *Another* to expect, who hath not these Advantages on his Side: tho' he should be found armed with unspotted Integrity, and unequall'd Learning, and Judgement?
>
> Your Holiness will judge from hence, how the Matter of *Heresy*

> stands amongst us: and how it must stand, unless my Lords the Bishops, who have with an unexampled Courage preserved our Liberties in Civil Matters, with equal Resolution step in; and oppose that Spirit, which from such Beginnings at first amongst you, proceeded farther and farther, till it broke out into Fire and Massacre, for God's Glory, and the Good of his Church.[104]

Samuel Richardson's drawing of Whiston walking away from a crowd of his contemporaries in 1748 at Tunbridge Wells is perhaps not accidental (see Figure 2). Whiston's writings go very much against the current of the Enlightenment as he attempts to give a Newtonian reading of Scripture reconcilable at once with Newtonian science, natural religion, antitrinitarianism, Whiggery, and radical millennialism.

2

WHISTON, THE BURNET CONTROVERSY, AND NEWTONIAN BIBLICAL INTERPRETATION

In this chapter, I begin to build the case for my thesis that Whiston's writings reflect a truly Newtonian method of biblical exegesis. In Section I, I trace the historical background of Whiston's first book, *A New Theory of the Earth* (1696), in the context of the Burnet controversy. Burnet's *Theory of the Earth* (1681–90) and his *Archaeologiae Philosophicae* (1692) were immediately adapted to the deist cause of undermining the authority of scripture and the special providence of God supposedly revealed in them.

In his collection of deistic tracts entitled *The Oracles of Reason* (1693), Charles Blount published an English translation of those parts of Burnet's *Archaeologiae Philosophicae* that were especially suitable to the deistic project of ridiculing revelation. Although Burnet claimed that his work was "far from excluding Divine Providence, either ordinary or extraordinary," Blount nevertheless utilized Burnet's work to argue in favor of the generally provident creator-architect of deistic natural religion while dismissing the specially provident (miracle-working, prophecy-fulfilling, prayer-answering) God of the Bible as mere fabulous allegory.

In Section II, I show how Whiston attempted to meet this assault on the authority of Scripture. He delineated a method of interpreting Scripture that showed how the Mosaic account of creation makes good, literal, historically determinate sense and must not be dismissed as a mere fable. To vindicate Mosaic history, he used his knowledge of Newtonian mechanical principles and, especially, a methodological assumption that what Moses describes as the history of creation is simply what an ordinary observer would have witnessed had one been present at the time. Whiston summarizes his method in three *postulata* at the end of his lengthy introduction to *A New Theory of the Earth,* the "Discourse Concerning the Genuine Nature, Stile, and Extent of the Mosaick History of the Creation." To show that this method is essentially Newtonian, in the sense that Newton also held these principles, is the crux of this section. I do not mean to imply that Whiston's Newtonian method of interpreting Scripture is a method unique to Newton. Rather, I propose only that Whiston's method, announced in his introduction to *A New Theory* on "the Nature, Stile, and Extent" of Moses' account and stipulated in his *postulates,* is an attempt to rationalize Scripture for a new age, and that, in the main, Newton agreed with this method.

The evidence for this claim is that fifteen years before the publication of Whiston's *New Theory,* Newton, in private correspondence with Burnet about the proper approach to the interpretation of Genesis, had enunciated the bare bones of Whiston's more fully elaborated methodology. This fact, coupled with the manner in which Newton promoted Whiston's academic career following the publication of *A New Theory* and recommended that Whiston apply these same principles to the interpretation of historically fulfilled prophecies (in Whiston's 1707 Boyle Lectures) suggests that Whiston's basic methodological approach to the interpretation of Scripture is justifiably labeled "Newtonian."

Finally, I show in the last part of Section II how Whiston attempts to use his Newtonian method to show that both the specially provident God revealed in properly interpreted Scripture, as well as the generally provident architect-creator inferred in the design argument of natural religion, are valid concepts. Whiston, the Newtonian exegete, insists that the generally provident God of natural religion is perfectly consistent with the specially provident God of Scripture and uses his Newtonian method to show that this is the case.

In Section III, I conclude this chapter by analyzing Michael Hunter's claim that "Newtonianism" is too inchoate a doctrine to justify such a label, in the light of the Newtonian John Keill's interpretative methodology, which contradicts Whiston – and, I argue, Newton.

I. The background to Whiston's "New Theory of the Earth": the deist attack on the authenticity of revelation

In 1696, Whiston published a best-seller that received widespread critical acclaim. Between 1696 and 1755, Whiston's first book, *A New Theory of the Earth, From its Original, to the Consummation of all Things. Wherein The Creation of the World in Six Days, The Universal Deluge, And the General Conflagration, As laid down in the Holy Scriptures, Are shewn to be perfectly agreeable to Reason and Philosophy,* went through six editions.

As the title suggests, this work was intended to answer those who directly attacked the historicity of the Mosaic account of creation and who indirectly attacked the concept of the providential nature of God's relationship to mankind as revealed in Scripture. Whiston's chief opponents in this controversy were Thomas Burnet and critical deists such as Charles

Blount, who adapted Burnet's work for the destruction of revelation. However, the seventeenth-century rational undermining of revelation and of God's providential involvement in his creation began on the Continent and was clearly implicit in Descartes's cosmogony.[1]

Descartes's conception of the evolution of the structure of the universe as the result of material particles turbulently whirling in a plenum strikes at the foundation of the traditional conception of the natural world as the theater for God's providential action in the affairs of men, while simultaneously simply setting aside the biblical account. God's providential role in the Cartesian vortexes is reduced to that of creator-mechanic. Descartes's contemporary, Pascal, notes the antiscriptural, unchristian effect of this speculation upon the concept of God: "Je ne pardonner à Descartes: il aurait bien voulu, dans toute sa philosophie, pouvoir se passer de Dieu; mais il n'a pu s'empecher de lui faire donner une chiquenaude, pour mettre le monde en mouvement; après cela, il n'a plus que faire de Dieu."[2]

In 1662, the year of Pascal's death, another Christian apologist, Edward Stillingfleet, an Anglican divine and future bishop of Worcester, concerned himself with the possibility that Descartes's speculation was apt to be utilized as a foundation for atheism "by persons *Atheistically* disposed, because of his ascribing so much to the power of *matter*." Stillingfleet argues that Descartes's hypothesis cannot be the foundation of atheism, because of the necessity in Descartes's hypothesis for God to create matter and then to set it in motion. Furthermore, Stillingfleet observes of Descartes "that this great improver and discoverer of the *Mechanical power of matter*" also maintains that God conserves and upholds the motion of matter, and so is definitely not atheistic. Stillingfleet never addresses the question of whether Descartes's vortex theory, "the particular *manner,* which was used by *God* as the *efficient cause* in giving being to the world," is an adequate account nor how it would conform with the Mosaic description if it were.[3] This is the task undertaken by Thomas Burnet.

Burnet (ca. 1635–1715), master of Clare Hall, fellow and proctor of Christ's College, Cambridge, chaplain to King William, master of the Charterhouse, and candidate for archbishop of Canterbury following John Tillotson's death in 1694, saw the danger of Cartesian mechanism for revelation and for the Christian providential deity. His work, first and foremost, is "an apology for and exposition of the idea of Providence in an age increasingly dominated by a mechanism and scepticism."[4]

Burnet's most famous effort to account for earth history – from the creation, through the Noachian Deluge, to the final conflagration – in a way that retained both the providential deity of the Bible and the mechanistic principles of Cartesian cosmogony is his *Telluris Theoria Sacra.* This book appeared in Latin in two sets of two volumes, the first set being published in 1681 (translated as *Theory of the Earth,* in 1684) and the second set in 1689 (translated in 1690). By 1700, this work had stimulated thirty replies, including Whiston's *New Theory of the Earth.*

Burnet's theory of the formation of the earth is firmly grounded in the theoretical speculations of Cartesian cosmogony. Like Descartes, Burnet believes that the planets of the solar system are former stars that, having their central fire (composed of the primary Cartesian element) encrusted with a layer of the dense, opaque tertiary matter, wander from vortex to vortex. Comets "like shadows wander up and down through the various Regions of the Heavens, till they have found out fit places for their residence, which having pitched upon they stop their irregular course, and being turned into Planets move Circularly about some Star."[5]

After this beginning, the earth passes through successive stages of "mechanical" development. As with "an eminent Philosopher of this Age, *Monsieur des Cartes,*"[6] Burnet's theory culminates with the rupture of the earth's crust and the formation, from the smooth original, of an earth deformed by jagged mountains, unarable plains, and irregularly shaped seas.

Burnet delighted in rationalistic theorizing about the parts of the "Frame of Nature":

> To make the Scenes of Natural Providence considerable, and the knowledge of them satisfactory to the Mind, we must take a true Philosophy, or the true principles that govern Nature, which are Geometrical and Mechanical. By these you discover the footsteps of the Divine Art and Wisdom, and trace the progress of Nature step by step.[7]

After tracing step-by-step the mechanical progress of the earth from a retired comet through the stages of its natural development, which culminates with the collapse of the crust, the ensuing Deluge, and the reduction of the pristine original earth to its present "great Ruine," Burnet explains how he has avoided the seemingly pure, and purely theoretical, mechanism of Descartes. Burnet claims that his theory is "far from excluding Divine Providence, either ordinary or extraordinary, from the causes and conduct of the Deluge." Rather than opposing God's providence as revealed in Scripture, Burnet argues that

> a disposition or establishment of second causes as will in the best order, and for a long succession, produce the most regular effects, assisted only with the ordinary concourse of the first cause, is a greater argument of wisdom and contrivance, than such a disposition of causes as will not in so good an order, or for so long a time produce regular effects, without an extraordinary concourse and interposition of the First Cause.[8]

Burnet's enthusiasm for "mechanical" explanations of natural events that are "satisfactory" to the mind leads him at this point to conflate two distinct categories of divine providence. God's providence, in the traditional Christian sense, consists of both a general providence and a special providence. Under the category of general providence is God's ability to create matter and the laws of nature. Because Descartes himself insists on this role for God, Stillingfleet had ruled out the possibility that Descartes was an atheist (see note 3). But there is a second sense of providence, special providence, of more importance to people such as Pascal who are distressed by the conception that the creation of matter and its laws is all that God had done (see note 2). Christians of the heart, such as Pascal, do not deny the nature of God's generally provident act of creation, but they insist that in addition to this generally provident act God remains active within his created world of natural law at least by preserving the frame of nature from destruction, and, usually, by answering prayers and forgiving secret sins. Occasionally God intervenes in what Burnet styles the "ordinary concourse" of *natura naturans* to produce a miracle. Such an "extraordinary concourse" on the part of God is special providence. It implies at least a daily superintendence of his generally provident creation, much as a computer programmer who has fed the machine its instructions might sit back and observe the printout of the programmed results. It often implies the preservation of the world of natural laws. But it also implies the real potency or power to break the natural order of natural law with a transcendent act that cancels the computer program and, perhaps, throws the whole machine out. The Bible, Burnet thought, records many instances that illustrate both sorts of providence.

Burnet pays lip service to God's special providence. "I know," he says, "a Sparrow doth not fall to the ground without the will of our Heavenly Father, much less doth the great World fall in pieces without his good pleasure and superintendancy." But it is clear that for Burnet, the "ordinary concourse" of nature is on so vast a scale and is so perfectly synchronized with the events of human history that it itself is an act of miraculous special providence. Burnet cautions his readers not to misconstrue his theory. Do not read it, he urges, "as if this Theory, by explaining the Deluge in a natural way, or by natural causes, did detract from the power of God, by which that great judgement was brought upon the World in a Providential and miraculous manner."[9]

The causing of the Flood by the natural, mechanical progression of the earth to a stage where the crust shattered was an act of special providence, an "extraordinary effect." Further evidence for the mammoth scale of God's providential superintendence of the earth is provided by the synchronization of the generally provident evolution of the earth in accordance with natural law and with the progressive moral degradation of the moral nature of mankind. Man had sinned sufficiently, by the precise time when the earth's crust was shattered by the mechanism of the sun's "sucking out the moisture that was the cement of its parts," to merit a just punishment. God's "ordinary concourse," his generally provident original design, perfectly synchronizes with the world of created nature and of man's moral nature. Burnet thus argues that

> it is no detraction from Divine Providence, that the course of Nature is exact and regular, and that even in its greatest changes and revolutions it should still conspire and be prepar'd to answer the ends and purposes of the Divine Will in reference to the *Moral* World. This seems to me the great Art of Divine Providence, so to adjust the two worlds, Humane and Natural, Material and Intellectual, as seeing through the possibilities and futuritions of each, according to the first state and circumstances he puts them under, they should all along correspond and fit one another, and especially in their great Crises and Periods.[10]

Burnet's conflation of special and general providence, or rather his elimination of special providence, in the traditional sense, from God's nature, has no repercussions for the biblical account of the Deluge. Burnet spins out his theory until all the effects of the present earth are explained in natural terms. Unlike Descartes, he thus reconciles some of his speculations with the "matters of fact" in the Bible. He is proud that his theory of the Flood, with the waters flowing out from the "Abysse" beneath the ruptured crust[11] and the subsequent reduction of human life spans to adapt to the new harshness of the seasons,[12] conforms to the letter of "*Moses's* Narrative of the Deluge . . . both as to the matter and manner of it."[13]

The great success of Burnet's *Theory of the Earth,* the English translation of which had gone through six editions by 1726 and provoked numerous responses, helped to create and exacerbate the problem that Whiston's *New Theory of the Earth* is designed to answer. Whiston studied Burnet's work, and Burnet's *Theory* became the subject of Whiston's examination for his first university degree. He remained favorably disposed to Burnet's *Theory* until his

> deeper researches into Mechanical Philosophy and the discoveries contained in Mr. Newton's wonderful book began to convince me of the indefensibleness of many of the particulars; and that the whole scheme as it then lay, could not be justified by the Principles of sound philosophy, nor did it, upon better consideration agree with the accounts of Holy Scripture.[14]

While satisfied that only his theory of the Deluge is consistent with Mosaic history, natural "phaenomena," and physical theory,[15] Burnet points out in 1692, in his own *Review* of his *Theory of the Earth,* that an objection, indeed a "principal objection," to his theory may arise from the apparent conflict between his Cartesian cosmogony of the earth's original creation and subsequent development *and* the Mosaic history of creation in six days as described in Genesis.[16] Indeed, Burnet dismisses the Mosaic history of creation in six days as an allegorical account that Moses "suited to the capacity of the people."[17] For Burnet, it is a "great question" whether the Mosaic history of creation "was ever intended for a true Physical account of the origine of the Earth: or whether *Moses* did either Philosophize or Astronomize in that description." The Mosaic history is reconcilable with Burnet's Cartesian cosmogony only "if we suppose it writ in a Vulgar style, and to the conceptions of the people."[18]

This is strong stuff, especially from a man considered by many to be in line for the job of Archbishop of Canterbury. In his next book, *Archaeologiae Philosophicae* (1692), he goes further. The many doubts that Burnet expressed in this work about the Mosaic creation story were translated into English by Henry Brown and published by the infamous deist Charles Blount in his notorious *Oracles of Reason* in 1693. Blount was an advocate of Herbert of Cherbury's natural religion and thought God's general providence was the only sort there was. He was one of the first to publicize Spinoza's view of miracles as simply a word that ignorant people apply to an event they do not understand but that really happened naturally, since, in Spinoza's pantheistic theory, God *is* nature.[19] Blount quotes Bishop Sprat's argument, in his *History of the Royal Society,* that God's general providence as an adept creator-designer is better demonstrated by a strong and inviolable natural order than by transgressions of that order in miraculous acts of special providence.[20] In an earlier work, Blount had publicized and endorsed Hobbes's view that established religion, with its "revealed" accounts of miraculous acts of divine special providence, originated in human superstition and ignorance, a view adopted in the opening lines of Spinoza's preface to his *Tractatus.*[21]

For his assault on the authority of revelation, Blount happily makes use of Burnet's careful arguments that although the hexameron and the Cartesian theory both suppose an original chaos, beyond this bare and very general agreement, "if we seek only after pure naked Truth, and a Physical Theory," we must abandon Moses as a vulgar adaptation for people "uncapable of being made philosophers."[22] Blount set down Burnet's doubts about the literal accuracy of Moses' account in a deistic tract in which he also espoused a Brunoesque eternity of a plurality of worlds, the materiality of the soul, the denial of original sin and miracles, and even doubt that Moses wrote the Pentateuch.[23]

The danger of Burnet's position, as Blount delightedly perceives, is not that it casts doubt on the generally provident, sacred origins of the earth and the whole frame of nature but that it calls into question the sacred origin of Scripture itself. Deists such as Blount and, in the next century, John Toland, Anthony Collins, and even David Hume, accept the design argument to some degree as illustrative of a generally provident deity of some sort. What they reject is the revealed religion of the Bible, with its "histories" of miraculous acts of special providence. Throughout his career, and especially in *The Oracles of Reason,* Blount repeatedly raised the ultimate skeptical problem about the divine origin of Scripture. In this project, Burnet's work was naturally coopted by Blount. Burnet's continual emphasis on the mechanical, secondary causes of divine general providence and his skepticism about Moses' creation story are naturally suited to the deist attempt to set aside revealed religion in favor of "natural religion" and the corollary attempt to eliminate the specially provident, universe-preserving, miracle-working, prayer-answering deity there revealed.[24] Blount's adaptation of Burnet's arguments for this purpose is encapsulated in a popular ballad of the period in which "Burnet" is described as saying,

> That all the books of Moses Were nothing but supposes.[/] That as for Father Adam and Mrs. Eve, his Madame,[/] And what the devil spoke, Sir,[/] 'Twas nothing but a joke, Sir,[/] And well-invented flam.[25]

Here, then, is the problem that so much of Whiston's effort is meant to solve. Whiston first attempts, as we will see in the following sections of this chapter, to vindicate revelation from such doubts about its divine origin and sacred authority. I argue that his attempt is distinctly Newtonian. Second, as I show in Chapter 3, he attempts to use properly interpreted Scriptural revelation to illustrate divine special providence.

Blount's use of Burnet's arguments and his "vindication" of Burnet's *Archaeologiae Philosophicae*[26] undoubtedly hurt Burnet's chances of succeeding his mentor Archbishop Tillotson as Archbishop of Canterbury when the latter died in 1694. Many bishops believed "that some of his Writings were too Sceptical,"[27] and Thomas Tenison became the next archbishop. Burnet retired to the Charterhouse. With his *Theory of the Earth* and its extrapolation of Descartes's cosmogony through the successive developments in the earth's natural history, Burnet had helped raise one of the crucial skeptical issues of the eighteenth century – the question of how to demonstrate the divine nature of revelation. William Whiston, among others, attempted to answer Burnet, but in a particularly "Newtonian" fashion.

II. Whiston's "Newtonian" response to the perceived deist threat: validating revelation

Whiston's New Theory of the Earth

For Burnet, the "hinges upon which the Providence of this Earth moves" are best revealed in his extrapolation of the cosmogonic centrifuge of Cartesian vortexes, not in the revelation given to Moses.[28] For Whiston, the dangerous ammunition Burnet provides to deists such as Blount is sufficient reason to expose this radically destructive view of Scripture by developing an explicitly Newtonian method of scriptural interpretation that harmonizes scriptural accounts of creation and the Flood with Newtonian physical science. He aims his work directly at Burnet, who, he says, has made the "unhappy Slip" of asserting Mosaic history "to be a meer Popular, Parabolick, or Mythological relation," with disastrous consequences.[29] Whiston excuses Burnet, an "excellent Author," from intending to help the deist cause but observes morosely the sad progress of the "contemners" of Scripture as a result:

> 'Tis well known how far this matter has been carried by Wise and Good Men; even to the taking away the literal, and the resolving the whole into a Popular Moral or Parabolick sense: And under what notion this History on the same account has appear'd to others, of no less free, but less Religious Dispositions and Thoughts, I need not say: What is indeed matter of doubt and perplexity to pious men, being unquestionably to the Loose and Profane, the Subject of Mirth and Drollery, and the sure encouragement to Atheism and Impiety.[30]

Burnet had gone too far by interpreting the Mosaic creation story as "no more to be accounted for or believ'd than the fabulous representations of *Aesop*." To vindicate revelation requires that the exegete have both "a solid acquaintance with, (not ingenious and precarious Hypotheses, but) true and demonstrable principles of Philosophy" and a similar understanding of "the History of Nature, and with such ancient Traditions as in all probability were deriv'd from Noah."[31] Newton's natural philosophy is the principal "medium" for buttressing revealed religion against deism. Science, history, and textual scholarship all combine in Whiston's brand of Newtonianism, not as a haphazard mixture but as a coherent attempt to secure revelation from deist ridicule. To counteract their calumny, what is required is neither simpleminded literalism nor sermon invective but a clear and rational comparison of the details of Newtonian science with revealed history. Whiston counsels, in his Newtonian method of exegesis, that

> wise men would rather set themselves carefully to compare Nature with Scripture, and make a free Enquiry into the certain *Phaenomena* of the one, and the genuin Sense of the other; which if Expositors would do, 'twere not hard to demonstrate in several such cases, that the latter is so far from opposing the truths deducible from the former . . . that 'tis in the greatest harmony therewith.[32]

The problem, of course, is the method of deriving the "genuin Sense" of Scripture. Whiston proposes a way to interpret Moses that "shall at once keep sufficiently close to the Letter of *Moses,* and yet be far from allowing what contradicts Divine Wisdom, Common Reason, or Philosophick Deduction."[33] Whiston (and Newton) agree completely with Burnet's point that Moses gears his account to the capacity of his "vulgar" audience. The Mosaic creation story is not a "Nice and Philosophical account of the Origin of All Things."[34] For Whiston (and Newton), as well as Burnet, what counts as a "Nice and Philosophical" explanation of the phenomena of creation is a mechanical description of their natural, secondary causes. In Whiston's *New Theory,* comets cause the major transformations of the creation, the earth's diurnal motion, and the Deluge. Unlike Burnet, however, Whiston insists that although Mosaic history is not a precise "mechanical" explanation, neither is it a fable. It is "an Historical and True Representation of the formation of our single Earth out of a confus'd Chaos, and of the successive and visible changes each day, till it became the habitation of Mankind."[35]

This seminal distinction between "historical" description and "scientific explanation" is the rock upon which Whiston erects his Newtonian method of biblical interpretation. The distinction is subtle. The revealed word of God, delivered through Moses, is true in that it describes historical events as they would have appeared to an observer on the scene; thus, it is a legitimate source for the Newtonian exegete. But revealed history does not provide a scientific explanation of the events described, though it is capable of being interpreted so that it harmonizes with the new astronomy and Newtonian mechanics. By distinguishing what one can obtain from revelation (a historically accurate description) and what one cannot (a scientific explanation of events that yet is compatible with the literal description), Whiston seeks to thread his way between the "wildness and unreasonableness" of those simpleminded literalists who accept revelation as both a historical description and a scientific explanation and those "loose Deists" such as Blount who utilize Thomas Burnet's honest doubts about literally interpreted Mosaic history to assert that revelation is neither historically accurate or conformable to science, and hence no more to be believed than Aesop.

Whiston grounds his scientific explanations in the well-developed seventeenth-century English tradition of probabilism, which has been traced by Henry G. Van Leeuwen.[36] " 'Tis evident," Whiston says, "that all Truths are not capable of the same degree of evidence, or manner of Probation."[37]

The most firmly established conclusions are "First Notions Known by Intuition" – for example, the Being of God.[38] Our assent to such a priori, self-evident propositions is so quick that we can scarcely observe any process of ratiocination.

"Purely Mathematical Propositions," Whiston's second category, "are demonstrated by a chain of deductions, each of which is certain and unquestionable."[39] Each link in such a chain of argumentation is built of a priori, self-evident propositions and so is known just as infallibly as any first notion.

The third category are propositions that are arrived at by a combination of purely abstract mathematical chains of argumentation and observations of natural phenomena that provide the links in the chain. Connecting the links of observed phenomena is done purely mathematically, but the links themselves, in such a mixed "probation," are empirically derived by observation, and so the level of certainty is less than that possible in the first two categories. The truth of optical, astronomical, and geographical propositions, " 'tho not arriving to the strict infallibility of the former sort, are yet justly in most cases allow'd to be truly certain and indubitable."[40]

The fourth category of "probation" is straightforward Newtonian hypothetico-deductivism. This category comprises knowledge of the truth of propositions (or "hypotheses") that state the physical causes of effects empirically observed in the world of nature. Such propositions are considered to be demonstrated when the effects observed necessarily "depend on certain supposed causes . . . with the impossibility of producing an instance to the contrary" and when they are inexplicable on the basis of "any other Hypothesis." Whiston states that

> this last method is that which our best of Philosophers has taken in his Demonstration of . . . *Gravitation,* and which accordingly he has establish'd beyond possibility of contradiction; and this is the sole way of bringing natural knowledge to perfection, and extricating it from the little *Hypotheses,* which in defect of true Science, the World has till lately been forc'd to be contented with.[41]

It is this last method, with one qualification, that Whiston uses to "demonstrate" his various hypotheses regarding the natural phenomena of earth history. Whiston states that

> ... where a Cause is assigned, whose certain consequent Effects must be very many, very surprizing, otherwise unaccountable, correspondent on the greatest niceness of Calculation in the particular Quantity and Proportion of every Effect, and where withal no disagreeing *Phaenomena* can be urg'd to the contrary; the evidence hence deriv'd of the reality of the assigned Cause, tho' of a different nature, and, if you will, degree too, from Demonstration, is yet little less satisfactory to the minds of wise and considering Men, than what is esteem'd more strictly so.[42]

But Whiston's method is not strictly a hypothetico-deductive one, and the reason why it is not further lessens the degree of certainty possible while dramatically underscoring the necessity of defeating the deist assault on the veridicality of Scripture history. The effects or "phaenomena" relating to the creation from a cometary chaos, the beginning of diurnal motion at the time of the Fall through the impact of another comet; and the Deluge, caused by the near miss of a third hypothesized comet are all based on scriptural accounts of those effects. For Whiston, there is a fifth category of "probation," that of propositions relating to events and phenomena derived from history. Whiston says, "History is all that we commonly can have for matters of fact past and gone; and where 'tis agreed upon by all, and uncontrollable, 'tis esteemed fully satisfactory, tho' not absolutely certain in common Cases."[43]

Whiston's Newtonian hypothesizing about mechanical causes of historical events rests on the foundation of the historical reports of those events, and so the certainty of the basic hypothetico-deductive method, which is itself almost but not quite absolute, is further reduced. Nevertheless, it results in a degree of certainty, Whiston says, "as satisfactory as the Nature of the thing is capable of." For example, if one assumes that a comet was the cause of the Deluge, the effects related in the Mosaic account are almost certain, or at least very probable.[44]

A further probabilistic feature of Whiston's historical hypothetico-deductivism is that he does not expect it to be able to resolve every single problem of exegesis. We ought not even to expect such comprehensiveness:

> Our understandings are finite, our Capacities small, our Sphere of Knowledge not great. We depend on God Almighty as to what we *know*, as well as what we *have,* or what we *are*. 'Tis possible it may not yet be the proper season for unravelling the Mystery, and so the requisite helps not yet afforded: our own unskilfulness or prejudices; some false Notion or precarious *Hypothesis* we have embrac'd; our misunderstanding the

> Nature of the Scripture Stile; a Mistake of a Copy; the ignorance of the various Stages and Periods of the World to which the particulars belong . . . may justly be supposed the occasions of our difficulties, without calling in question either the truth of our Humane Faculties, the Attributes of God, the *Phaenomena* of Nature, or the genuine sense of Holy Scriptures.[45]

For Whiston, "Knots of Scripture and Providence" may be beyond our rational comprehension, but they can never be directly contradictory to it. In any case, such instances are few, and the Mosaic creation story is not one of them.

Whiston saves the literal, historical accuracy of Moses' hexameron while rejecting it as a scientific explanation of the causes of those events, or as a "Nice and Philosophical Account," by arguing that the descriptions of Moses, with the exception of the account of the original creation of all of matter *ex nihilo* (Gen. 1:1),[46] apply only to the formation of this earth, and not the entire "vast Universe." Only the vulgar could believe that this earth is "the only Darling of Nature, the prime Object of the Creation and Providence of God."[47]

Once Whiston has shown us that Moses is describing only the formation of the earth, he goes through Moses's "*Historical Journal or Diary of the Mutations of the Chaos, and of the visible Works of each day*,"[48] which any observer who had been present would have seen, to provide a scientific explanation. " 'Tis very reasonable to believe" that Moses' statement that the earth "without form and void" (Gen. 1:2) is a literally accurate description of the "atmosphere" of a comet.[49] With his Newtonian method of exegesis assuring him of the literally accurate historical data, Whiston demonstrates how "Mosaick history" harmonizes with the nature and trajectories of comets.

Whiston hypothesizes that in case such a "wild and disturbed Confusion" of particulate matter became a planet, – that is,

> if its Eccentrical *Ellipsis* were turn'd into a Concentrical Circle, or an *Ellipsis* not much differing therefrom; at a suitable and convenient distance from the Sun; there is no reason to doubt but the parts of that confused Atmosphere which now encompass it to such a prodigious distance, might subside and settle downwards according to their several Specifick Gravities; and both obtain and persevere as settled, fix'd, and orderly a Constitution as a Planet has.[50]

Whiston waffles about whether this change of orbit might have been produced by a miraculous intervention of special providence or whether it might rather have been programmed into the comet in question as part of God's generally provident creation. On the whole, he seems, with Burnet, to prefer the latter as demonstrative of a more clever and wise deity. He states that

> if we should suppose, as 'tis possible to do, that God did not by a miraculous Operation remove the *Chaos* or Comet from its very Eccentrick Ellipsis to that Circle in which it now began to revolve; but that he made use of the Attraction or Impulse of some other Body; yet in this case the Lines of each Bodies motion, the quantity of Force, the proper Distance from the *Sun* where, and the exact time when it happen'd . . . must have been so precisely and nicely adjusted before-hand by the Prescience and Providence of the Almighty, that here will be not a much less remarkable Demonstration of the *Wisdom, Contrivance, Care* and *Goodness,* than the other immediate Operation would have been of the *Power* of God in the World.[51]

It follows from this hypothesis of a change from a radically elliptical orbit to a nearly circular one that a day, at this time, was the length of a year according to our current measure, because although the chaos now moved annually around the sun, it had as yet no diurnal motion.[52] Because a year equaled a day in physical fact prior to the Flood, the historical accuracy of Moses' history that creation took six days is preserved. The six "days" of creation are six years.[53]

Seriatim, Whiston harmonizes Moses' literal description of the phenomena of the hexameron with this scientific explanation of those same phenomena. The "Spirt of God" moving "upon the face of the waters" (Gen. 1:2) refers to the settling out process of "the Mass of Dense Fluids, which compos'd the main bulk of the intire *Chaos*."[54] The creation of light and its separation into night and day (Gen. 1:3–5) refers to what would have been visible as the more "Earthy and Opake Masses," being denser, settled toward the center, admitting light from the sun into the more rarified upper atmosphere.[55] The making of the "firmament" and the separation of waters above and below the firmament (Gen. 1:6–8) refers to the continuation of the settling process as clouds and seas formed.[56] The appearance of the dry land (Gen. 1:9–10) refers to the conglomeration of solid earthy corpuscles upon the denser fluids of the abyss; by a version of isostasy, these formations sank according to their specific gravities. Mountains are higher than valleys because, being composed of less dense "columns" of matter, they subsided less.[57] Into these formations, the "superior" waters that had evaporated into the atmosphere fell as rain. This rainfall, which produced small seas, left the earth "moist and juicy" for the germination of the seeds of vegetables, and so Moses describes the production of vegetation on the third day (Gen. 1:11–12).[58] Whiston admits that the creation of seeds of both plants and animals was done by "the immediate

Workmanship of God," by which he means a generally provident work at the time of the original creation.[59] The "making" of the lights of the firmament on the fourth day (Gen 1:16) means to a scientist that "these Heavenly Bodies, which were in being before, but so to be wholly Strangers to a Spectator on Earth, were rendered visible."[60] The production of fish and fowl on the fifth day (Gen. 1:20–3) proceeded naturally from the "Seeds, or little Bodies of Fish and Fowl which were contain'd in the Water" from the original act of creative, general providence.[61]

The sixth day's work, the creation of man was, however, a genuine act of special providence – a miracle:

> Tho' 'tis granted that all the other Day's Works mentioned by *Moses* were generally brought to pass in a natural way by proper and suitable Instruments and a Mechanical Process, as we have seen through the whole *Series* of the foregoing Creation; yet 'tis evident . . . That an immediate and miraculous Power was exercis'd in the formation of the Body, and Infusion of the Soul of Man.[62]

In between the generally provident act of creation and the specially provident, miraculous interposition of God in his creation on the sixth day with the creation of man, everything described by Moses happened naturally. But what he is describing is, in all probability, explained or interpreted as the result of the change of a cometary chaos from a radically elliptical orbit to a circular, or almost circular, one and its subsequent natural development. Whiston thus adheres to one of the principal postulates of his method of interpretation: "That which is clearly accountable in a natural way, is not, without reason, to be ascrib'd to a Miraculous Power."[63]

Whiston attempts to avoid Blount's deistic reading of Burnet in which Scripture is reduced to the status of a childish fairy tale, while also avoiding the simpleminded literalism of fanatics oblivious to the dictates of scientific reason. He preserves the Mosaic hexameron as a historically accurate description – by no means in the category of Aesop's fables – while simultaneously interpreting it in a manner acceptable to a Newtonian scientist of the early eighteenth century. He does not stop with the creation. He goes on to explain that after mankind's Fall from the state of innocence came God's swift punishment of the Fall in the Noachian Deluge, accomplished with the mechanism of comets.

In paradise, the air was warm, and free of winds and storms. Lacking diurnal rotation, the primitive earth was perfectly spherical. Its orbit was very moderately elliptical. There were no seasons. And, of course, in this paradise, humanity was innocent.[64] But then Adam and Eve sinned, and Whiston hypothesizes that with the impact of a comet "obliquely upon the Earth along some parts of its present equator"[65] would explain many changes in the earth, as well as the mechanism and nature of the divine punishment. The comet's impact caused the end of paradise: the earth's shape changed from that of a perfect sphere to that of an oblate spheroid; the earth's orbit changed from a slightly eccentric to a concentric one;[66] and the cometary impact tilted the earth's axis, producing seasons, winds, tides, and diurnal rotation. All of this, Whiston argues, agrees perfectly with the time and nature of the effects of God's punishment as described in the Bible, in contrast to Burnet's theory, in which the Fall is merely a long, evolutionary slide into decay that occurs over sixteen centuries, from the time of creation to that of the Flood. In God's adjustment of the mechanical causes of the natural world to the needs of the moral world, we are presented with further evidence of the comprehensive power of God's general providence at the creation. For Whiston, the natural and mechanical conjunction of the course of the comet with the precise moment when it was necessary to punish excessive wickedness is

> the Secret of the Divine Providence in the Government of the World, and that whereby the Rewards and Punishments of God's Mercy and Justice are distributed to his Rational Creatures, without any disturbance of the settled Course of Nature, or a Miraculous interposition on every occasion. Our imperfection is such, that we can only act *pro re nata,* can never know before hand the Behavior or Actions of Men; neither can we foresee what Circumstances and Conjectures will happen at any certain time hereafter; and so we cannot provide for future Events, nor predispose things in such a manner that every one shall be dealth. . . . But in the Divine Operation 'tis quite otherwise: God's Prescience enables him to act after a more sublime manner; and by a constant Course of Nature, and Chain of Mechanical Causes, to do every thing so, *as it shall not be distinguishable from a particular Interposition of his Power*. . . . He who has created all things . . . At once looks through the intire Train of future Causes, Actions and Events, and sees at what Periods, and in what manner 'twill be necessary and expedient to bring about any changes, bestow any Mercies, or inflict any Punishments on the World: Which being unquestionably true, 'tis evident he can as well provide and predispose natural Causes for those Mutations, Mercies, or Judgements before-hand; he can as easily put the Machin into such Motions as shall, without a necessity of his mending or correcting it, correspond to all these foreseen Events or Actions, as make Way for such Alterations afterward. . . . And when these two ways are equally possible, I need not say which is most agreeable to the Divine Perfections and most worthy of God. So that when the Universal Course of Nature, with all the Powers and Effects thereof, were at first deriv'd from, and are continually upheld by God; and when nothing falls out any otherwise, or at any other time, than was determin'd

> by Divine Appointment in the Primitive Formation of the Universe: To assign *Physical* and *Mechanical* causes for the Deluge, or such mighty judgements of God upon the Wicked, is so far from taking away the Divine Providence therein, that it supposes and demonstrates its Interest in a more Noble, Wise and Divine manner than the bringing in always a miraculous Power wou'd do.[67]

The Noachian Deluge may be explained by the same scientific hypothesis of the effects of a comet. This one did not hit the earth but was a near miss. It accounts mechanically for the time, quantity, and circumstances described by Moses with great literal accuracy. The "opening of the windows of heaven" (Gen. 7:11) and the forty days and nights of rain (Gen. 7:12) refer to the effects of the passage of the earth through the comet's vaporous tail.[68] Moreover, since the Fall, when the earth's perfectly spherical shape had been altered to that of an oblate spheroid, the earth's crust, although it remained solid, and stretched and cracked. The weight of the rainfall from the comet's tail and the tremendous tides in the dense fluid of the Abyss caused by the gravitational attraction of the comet's mass cracked the crust, releasing the "fountains of the great deep" (Gen. 7:18).[69] Figure 3 illustrates how the close approach of a comet could produce these effects.

The retreat of these floodwaters took one or two hundred years and is confirmed by fossils.[70] In addition, the added mass of the floodwaters, as they evaporated or subsided underground, deposited a residue of earthy matter that retarded the annual orbit of the earth so that it now circled the sun once every three hundred sixty days. This fact is attested to by the calendars of many ancient civilizations.[71] Later Whiston attempted to demonstrate that Halley's comet had been the particular mechanical cause of the Deluge.[72] The unfulfilled prophecies of a general conflagration of the world can also be explained as the result of a hypothesized future cometary approach that will again displace the earth's orbit, bringing it nearer to the scorching sun.[73]

All of these scientifically satisfactory hypothetical explanations of the historical data recorded in Genesis rest on the three "postulata" that conclude Whiston's "Introductory Discourse" to the book:

> I. The Obvious or Literal Sense of Scripture is the True and Real One, where no evident Reason can be given to the contrary.
>
> II. That which is clearly accountable in a natural way, is not without reason to be ascrib'd to a Miraculous Power.
>
> III. What Ancient Tradition asserts of the constitution of Nature, or of the Origin and Primitive States of the World, is to be allow'd for True, where 'tis fully agreeable to Scripture, Reason, and Philosophy.[74]

The "Newtonianism" of A New Theory

Whiston claims that he showed the manuscript of *A New Theory* to Richard Bentley and Christopher Wren but that it was "chiefly laid before Sir *Isaac Newton* himself, on whose Principles it depended, and who well approved of it."[75]

The fixed critical response to identifying the Newtonian principles that are supposedly the foundation of the work has generally focused upon the mechanism of comets in producing changes on the earth, without explaining why this is especially Newtonian.[76] A recent exception to this general tendency is the analysis of Roy Porter, who emphasizes the mechanism of comets as the mechanical cause of all significantly catastrophic changes on the earth but also places them in a particular Newtonian context. Porter argues that Whiston's comets are "external bodies," and so related to Newton's passive theory of matter. Because Newton conceives of matter as hard, inert, and impenetrable, he views the earth as essentially passive. Because earthly matter is not possessed of force, motion, or any sort of hylarchic vitalism, the terrestrial economy of matter is supremely stable and necessarily requires an external mechanism, such as comets brought into the terrestrial system by gravity, to effect significant changes.[77]

But as important as are the mechanism of comets, the passive notion of matter, and the other Newtonian principles used by Whiston[78] for elucidating his Newtonianism, of equal importance is his "Introductory Discourse." For Whiston's *New Theory* to make sense in its own context, it requires both the hypothesis of the comets and a theory of biblical interpretation. There are three reasons why his method of biblical interpretation is as much a part of Newtonianism as the mechanical concept of cometary influence and the metaphysical concepts of passive matter and gravity. First, Newton outlines, in his correspondence with Burnet, a view parallel to that of Whiston on biblical interpretation and states that he thereby achieves the same "probable" degree of certainty that Whiston achieves. Second, Whiston claims to have learned to oppose allegorical interpretation of prophecies from Newton, and Newton suggested a "literal," nonallegorical interpretation of biblical prophecy as the topic of Whiston's 1707 course of Boyle Lectures. Third, after the publication of *A New Theory,* Whiston's career was clearly advanced by Newton.

The evidence for the basic similarity between Whiston's method of biblical interpretation stated in the "Introductory Discourse" to *A New Theory* in 1696 and that favored by Newton lies in the correspondence between Burnet and Newton in 1680 and 1681. In the first extant letter, dated January 13, 1681, Burnet presses Newton hard on the nature of Moses' account of chaos and the creation in six days. Burnet was replying to a letter from Newton in which Newton had upheld the basic historicity of the Mosaic history if properly interpreted. Burnet is guided by three principles of interpretation. First, the "Mosaical formation of ye world is noe physical reality."[79] Second, Moses does not give a scientific theory because the people simply would not have understood him. Third, Moses "gives a short ideal draught of a Terraqueous Earth rising from a Chaos, not according to the order of Nature and natural causes, but in yt order wch was most conceivable to ye people, & wherein they could easily imagine an Omnipotent power might forme it."[80] In short, the Mosaic history is neither a physically accurate description of what happened nor a scientific explanation. It is only "ideal, or if you will, morall,"[81] and its purpose is to impress upon his followers the concept of an omnipotent deity in a way suited to their "vulgar" understanding.

If one proceeds, as Burnet states that Newton does in his previous letter – that is, according to "ye necessity of adhering to Moses his Hexameron as a physical description,"[82] then Burnet has some difficult questions to ask Newton. Of the light "made" on the first day, Burnet asks, "Wt was yt pray? wt physical reality, where made or how? was it made out of ye Chaos as other things, in wt manner pray?"[83] As for the "formation" of the firmament on the second day, Moses declares both that the firmament was divided into terrestrial and celestial waters and that the firmament was to be the seat of the sun, moon, and stars; Burnet asks,

> Now I appeale to any man whether these 2 local properties bee not utterly inconsistent? to divide betwixt ye Caelest. & terrest. waters it must be far below ye Moon; & ye celest. waters must be supposd betwixt it & ye Moon; and to bee ye seat of ye Sun Moon & Stars it must bee not onely as high as ye Moon but as ye Sun, nay as ye fixt stars wch are at an immense distance above ye Sun. Therefore ye Firmament wth these properties can bee noe physical reality. and see you how is another day of ye 6 imployd upon noe physical reality.[84]

Moses describes the creation of all out of nothing in the opening lines; why does God then remake the sun, moon, and stars on the fourth day? The only possible conclusion is that the original chaos applies only to the earth, not to the entire universe. "From wch concession," Burnet states, "I would infer . . . yt ye distinction of 6 dayes in ye Mosaical formation of ye world is noe physical reality."[85] Finally, in his previous letter, Newton had broached the possibility that the first revolutions of the earth had been much longer and "yt ye first 6 revolutions or days might containe time enough for ye whole Creation."[86] Burnet quashes this attempt to save the historical accuracy of Moses. This "supposition," Burnet states, "looks pretty well at first; but unless you make ye first 6 dayes as long as 6 yeares or rather much longer, I cannot imagine yt they should be sufficient for the work."[87]

Whiston's *New Theory,* written approximately fifteen years later, addresses each of Burnet's queries. But so does Newton, in his own reply to Burnet, and his letter states, at least in germ, every one of Whiston's three "postulates" of interpretation. Furthermore, even though the mechanism of comets is not mentioned in Newton's letter, several of his other mechanical hypotheses that explain the literally accurate historical data scientifically are stated by Newton. The crucial similarity between Newton's letter and Whiston's book, however, are the basic principles of biblical interpretation. It is this similarity that makes Whiston's method genuinely Newtonian, in the sense that Newton privately held and applied the same method.

Newton is adamant that while the Mosaic account is not "philosophical" – that is, scientific – neither is it "feigned." Moses accurately described physical realities, "much after ye manner as one of ye vulgar would have been incline to do had he lived & seen ye whole series of wt Moses describes."[88] This reading certainly accords with Whiston's first postulate that "the Obvious or Literal Sense of Scripture is the True and Real one" (see note 35). Whiston would certainly have agreed entirely with Newton's statement that for Burnet blithely to declare that Moses described events "when there was no such thing done neither in reality nor in appearance . . . is something hard."[89]

As for Whiston's second postulate that naturalistic, mechanical explanations of physical events are preferable to the interposition of miraculous, special providence unless we have adequate reason to believe otherwise, Newton states the identical approach: "Where natural causes are at hand God uses them as instruments in his works, but I doe not think them alone sufficient for ye creation."[90] Nor does Whiston, for whom the original creation of matter and of humanity is, in fact, miraculous. Whiston's third postulate, about using ancient profane tradition to corroborate "Scripture, Reason, and Philosophy," as in the case of his use of ancient calendars to corroborate his hypothesis of a three hundred sixty-day year, is not stated by Newton. Yet Newton does rely on this postulate in this letter when he conjectures that "one may suppose that all ye planets about our Sun were created together, there being in *no history* any mention of new ones appearing or old ones ceasing."[91] These are "principles" of Newton on which *A New Theory* depends as much as on his physical or metaphysical principles. In addition to his statement of Whiston's three postulates of interpretation, the whole of Newton's reply to Burnet is suffused with Newton's cautious, hypothetical probabilism, just as Whiston's book is. Newton observes that his hypothesis that the formation of the mountains at the time of creation proceeded by coagulation in a manner analogous to the way saltpeter in solution coagulates irregularly into long bars (that is, by a process of settling out of the chaotic matter) is not more probable than Burnet's own hypothesis of the formation of mountains by the shattering of the earth's crust. Indeed, Newton remarks that Burnet's hypothesis may be more probable.[92] When he cautiously puts forward his scientific explanations of Moses' historical descriptions, he does so by way of "conjectures."[93]

Finally, although he does not refer to the mechanism of comets, many of Newton's conjectured scientific explanations of Moses' words of description are strikingly similar to those in Whiston's book published fifteen years later. For example, the work of the fourth day was not to create stars. Newton states that the stars only appeared visibly on the fourth day:

> I do not think their creation from beginning to end was done ye fourth day, nor in any one day of ye creation nor that Moses mentions their creation as they were physicall bodies in themselves some of them greater then this earth & perhaps habitable worlds, but only as they were lights to this earth, & therefore though their creation could not physically [be] assigned to any one day, yet being a part of ye sensible creation wch it was Moses's design to describe & it being his design to describe things in order according to ye succession of days alloting no more than one day to one thing, they were to be referred . . . rather to ye 4th day then any other if the air then first became clear enough for them to shine through it & so put on ye appearance of lights in the firmament to enlighten the earth.[94]

This hypothetical explanation is a precise parallel to that of Whiston, who later echoes Newton's hypothesis that mountains were the result of the differentiating process of the original chaos begun on the first day and finally appearing to a hypothetical observer on the second.[95] Similarly, Newton restates his view that the original "day" was longer but admits that he can come up with no hypothesis to explain "any sufficient natural cause of the earth's diurnal motion" and so ascribes it to divine special providence, whereas Whiston conjectures that a small cometary impact set the world spinning.[96]

It is difficult to make out from direct references Newton's later opinion of Whiston's hypothetical comets as the initial chaos, the cause of diurnal rotation, and the cause of the Flood. There are hints. In the second edition of the *Principia* (1713), Newton gives more, and more accurate, examples of the calculations of cometary orbits. He notes that because comets are in fact in orbit, they do pass through the solar system and approach the other solar bodies; and "so fixed stars . . . may be recruited by comets that fall upon them."[97]

David Gregory, a prominent Scottish Newtonian, later repeated Whiston's hypothesis of the effect of a comet in passing through the solar system. If a comet passes near a planet, he asserts, "It will so attract [the planet] that its Orbit will be chang'd . . . whence the Planet's Period will be chang'd."[98] In December of 1694, more than a year before the publication of Whiston's *New Theory,* Edmond Halley read two papers to the Royal Society in which he hypothetically explained the Flood as the result of the "Choc of a Comet."[99]

As intriguing as these similarities of the scientific details are, the core of Whiston's Newtonianism is his respect for the Mosaic hexameron as a historically accurate description, even if it is not the sort of explanation a trained Newtonian scientist-theologian would provide. This sort of theorizing could always be added later, and as science progressed even those difficult "knots" of interpretation would unravel. The scientific explanation will always harmonize with the Mosaic appearances and descriptions. Given the striking parallels between Whiston and Newton in their interpretative approaches to revelation, I believe that I am justified in my claim that it is as much this epistemological method of interpreting Moses that links Newton and Whiston as the hypothesis of cometary mechanisms. That this kind of Newtonianism, which aims to reconcile Moses with science, was shared by Newton seems probable from Newton's like-minded letters to Burnet on how to read Genesis and his decisive intervention in Whiston's academic career following the publication of *A New Theory.*

Whiston's Astronomical Principles of Religion, Natural and Reveal'd *(1717)*

After publication of his *New Theory,* Whiston remained concerned with the threat of the modern infidels who had changed their names from "*Atheists* and acknowledged themselves to be *Deists*" – that is, those who pretended to believe and act in accordance with Scripture while subverting it by ridicule. But the moderns, by demonstrating God's creating and preserving general providence, had thwarted this attack. Whiston writes,

> The Elder Unbelievers never thought themselves safe in their Infidelity, till they had reason'd away God, or at least his Providence out of the World; and till they had advanc'd such Hypotheses in Philosophy as should account for the Make and Constitution of the World by bare Matter and Motion, without any allowance for a Divine Power therein: And the Modern Unbelievers have gone as far as ever they were able in the

> same way; disputing every Inch of Ground, where a Providence and the Power of God us'd to be suppos'd in the World. The Almighty has at last been pleas'd, by the noble Discoveries of late afforded us, to put an end to that Question, by letting us see that not one or two of the Phaenomena of Nature only, but almost all of them in general, are deriv'd from his immediate Divine Power and Influence.[100]

One of Whiston's most important demonstrations of God's providential control of the world is in his adaptation of the design argument to the Newtonian cosmology, in *Astronomical Principles of Religion, Natural and Reveal'd* (1717). But this work is far more than a classic illustration of how the new wine of the Newtonian system is poured into the old bottle of the design analogy to illustrate divine providence. Rather, Whiston's book, which is dedicated "to the illustrious Sir Isaac Newton, President, And to the Council and Members of the Royal Society," is aimed at demonstrating the adequate rational basis for believing "reveal'd," as well as "natural," religion. Whiston uses the design argument of Newtonian natural theology explicitly to support scriptural revelation. Whiston's *Astronomical Principles* is simply an extension of his defense of Scripture in *A New Theory,* and it is directed against the same deist enemies who denied the historicity of Scripture.[101] And, as with *A New Theory,* the central point of the *Astronomical Principles* is to show how to interpret the historical data preserved in Scripture; Whiston writes,

> Since it has now pleased God, as we have seen, to discover many noble and important Truths to us, by the Light of Nature, and the System of the World; as also, he has long discovered many more noble and important Truths by Revelation, in the Sacred Books; It cannot be now improper, to compare these two Divine Volumes, as I may well call them, together; in such Cases, I mean of Revelation, as relate to the Natural World, and wherein we may be assisted the better to judge by the Knowledge of the System of the Universe about us. For if those things containd in Scripture be true, and really deriv'd from the Author of Nature, we shall find them, in proper Cases, confirm'd by the System of the World' and the Frame of Nature will in some Degree, bear Witness to the Revelation.[102]

The Newtonian version of the design argument becomes, in Whiston's hands, an engine of destruction against deism, "Ill Mens *last Refuge.*" For Whiston, the design argument becomes part of a grand scheme of biblical criticism, and the astronomer, like the earth theorist, becomes a biblical interpreter who uses the truth about God revealed in the "Divine Volume" of Nature to check his understanding of the "Divine Volume" of revelation.

By comparing comments in Newton's letters with brief statements in his published works, from his letters to Bentley in 1692 to his "General Scholium" in 1713, the application of his system to the design argument for the existence of God can be traced. To Bentley, who first publicly propounded the Newtonian design argument in the inaugural Boyle Lectures of 1692, Newton writes that proving "beliefe of a Deity" on the basis of the "systeme" that he had demonstrated in the *Principia* had been present in his mind at the time of its writing.[103] By the "method of analysis,"[104] Newton argues,

> To make this systeme therefore with all its motion, required a Cause wch understood & compared together the quantities of matter in ye several bodies of ye Sun & Planets & ye gravitating powers resulting from thence, the several distances of the primary Planets from ye Sun & secondary ones from Saturn Jupiter & ye earth, & ye velocities wth wch these Planets could revolve at those distances about those quantities of matter in ye central bodies. And to compare & adjust all these things together in so great a variety of bodies argues that cause to be not blind & fortuitous, but very well skilled in Mechanicks & Geometry.[105]

The most famous example of Newton's design argument is the 1713 "General Scholium" to the second edition of the *Principia*. Here, too, Newton infers that the cause of the order empirically observed in the "most beautiful system of the sun, planets, and comets, could only proceed from the counsel and dominion of an *intelligent* and *powerful* Being."[106]

Whiston's *Astronomical Principles* is a textbook illustration of how a modern Newtonian scientist proceeds empirically to gather up the "effects" in nature, in order, by reasoning on the basis of analogy, to infer the nature of their cause. The first ninety pages of the book catalog Newton's "system," by far one of the most complete and detailed Newtonian explanations of the "appearances and phenomena" of the astronomical world. The format of the *Astronomical Principles* shows that Whiston considered the mechanical first principles of Newtonianism – the laws of matter and motion – to be certain. Thus the first chapter sets out as "lemmata" the "Known Laws of Matter and Motion."[107] Whiston next shows how the Newtonian laws of motion in the "lemmata" demonstrate "a universal *Power of Gravity* acting in the whole System." This power must exist, because the observed motions of the planets are curves, and so some power must perpetually draw them "from their natural Rectilinear Courses along strait Lines."[108] Whiston demonstrates that this power is proportional to the mass of the body it affects (the inverse square law) and then argues that it cannot possibly be a mechanical power because mechanical power or force is exerted only through external contact on the external surface of matter. The power of gravity, by contrast, acts upon the "very inward Substantial Parts of Bodies," and so, whatever its causes or nature, it is not any sort of mechanical pressure.

Whiston next observes that, given the "lemmata" of projectile motion, if the "immechanical" power of gravity were suspended for an instant, then "all the whole System would immediately dissolve; and each of the Heavenly Bodies would be crumbled into Dust; the single Atoms commencing their several Motions in . . . strait Lins."[109] Furthermore, there is no inherent necessity for the bodies of the solar system to require "that nice Adjustment . . . of the projectile Velocity to the Attractive Power through the whole Universe."[110]

After having presented what amounts to a popular course on Newtonian physics and astronomy, Whiston proceeds, reasoning by analogy, to harvest a rich yield: a proof of the existence and attributes of God. By analogy with human contrivances, it is apparent to Whiston that this whole system of nature is caused by the choice, prudence, and providential judgment of God. The process of inference is

> . . . the very same by which, from the Contemplation of a Building, we infer a Builder; and from the Elegancy and Usefulness of each Part, we gather he was a skilful Architect; or by which from the View of a Piece of Clockwork, we conclude the Being of the Clockmaker; and from the many regular Motions therein, we believe that he was a curious Artificer.[111]

Anyone who understands the principles of Newtonian physics and astronomy, he asserts, must also acknowledge the certainty of the inference that the divine artificer is not merely a generally provident creator-deity but also the specially provident preserver of this carefully contrived "System of the World." God exercises this continuing, superintending aspect of his special providence through the maintenance and preservation of the power of gravity. Without the continual exertion of this "immechanical" power, "all this beautiful *System* would fall to pieces."[112] In the *Astronomical Principles,* as in *A New Theory,* the routine progression of the laws of nature, including the law of gravity, becomes the supernatural, specially provident effect of God's continual act of preservation. Whiston says:

> I do not know whether the falling of a Stone to Earth ought not more truly to be esteem'd a *supernatural Effect,* or a *Miracle,* than what we with the greatest surprize should so stile, its remaining pendulous in the open Air; since the former requires an *active Influence* in the first Cause, while the latter supposes *non-Annihilation* only.[113]

Newton, in addition to his own statement of the inference of a generally provident creator "well skilled" in geometry and physics, also concurs that the creator fulfills a specially provident role by continually acting to preserve the mechanical system. For Newton as for Whiston, "A continual miracle is needed to prevent the sun and fixed stars from rushing together through gravity."[114] Whiston further specifies that this activity is the specially provident activity of a God still superintending his creation, because it reveals "a constant, uniform, active Influence or Energy in all the Operations done in it; the very same which was exerted in the Original Impression of those Laws of Motion on which it depends."[115] Whiston cites a line from his beloved *Apostolical Constitutions* that reveals again the agreement between the most ancient records and the explanations provided by Newtonian science. "*The Whole World*," by God's preserving special providence, is literally "*held together by the hand of God*."[116]

Whiston, like most other Newtonians, regarded the design argument, with its analogical inferences regarding the existence and nature of God, as demonstrably certain. Whiston finally extends the design argument to show how the absolutely certain knowledge thus demonstrated on the basis of the Newtonian "Divine Volume" of nature helps the Newtonian exegete to support the basic historicity of the other sacred book, the Bible. The Newtonian biblical interpreter has only to compare the two books, using the certain Newtonian system and inferences of natural religion as the criteria of truth. When the "Divine Volume" of revelation is found to conflict with the explanation of the "Frame of Nature, which is now much better understood than in the Days of those Antient Writers," then a forgery in the sacred volume of Scripture has been detected. Because the testimony of Scripture is the result of a long chain of human transmitters of God's original message, its level of certainty is below the inferences of the natural religionist, who bases his conclusions on the certain bedrock of the Newtonian frame of nature.

But, as with the earth theorist's naturalistic explanation of the Flood described by Moses, here, too, the "considering Astronomer" encounters no conflict between the certain inferences drawn on the basis of the frame of nature and those derived from revelation. In point of fact, the astronomer, like the geologist, discovers that the two "Divine Volumes" substantially confirm one another. As Whiston puts this point, the literal agreement of the certain inferences of natural religion with Scripture will

> ... be a mighty Evidence for the Truth, and Uncorruptness of those Scriptures; and this even in general, as to such other Contents of the same, as can no way come under the like Methods of Examination. If I am once fully satisfy'd, that a Witness is Upright and Honest, even in

several Points where there was the greatest Suspicion as to his Sincerity, he will deserve the better Credit in other Cases, even where no corroborating Evidence can be alleg'd for his Justification. To this Kind of Evidence then do I appeal on behalf of those Sacred Writings.[117]

Thus, in the section of the *Astronomical Principles* that he devoted to listing "Important Principles of Divine Revelation, confirm'd from the foregoing Principles, and Conjectures,"[118] Whiston claims only a high degree of probability in his argument. Just as Newton moves from the creating, preserving God (inferred on the basis of the analogy of nature with a human contrivance) to the Lord God of Israel,[119] so Whiston argues that "the Sacred Accounts declare the very same, *all the same Truths,* ascribe the very same, *all the same Attributes,* to God, which we have shewed to be real, and to belong to God, from the Consideration, of the True System of the World."[120]

Whiston is persuaded that Scripture records the same "truths" about and "attributes" of God as the inferences of natural theology do, since the testimony of Scripture is reliable, even though it will never be infallible in the way that mathematical deductions are. In *Astronomical Principles,* Whiston recommends an approach to interpreting the testimony of revelation that expands his brief remarks about the method of historical "probation" in *A New Theory*. He suggests that one approach the testimony of revelation in the same way that judges in English law courts adjudicate legal battles.

Even though mathematical certainty is not derivable from the human testimony regarding disputed affairs in life, judges do not retreat into the perpetual suspended judgment of "scepticalness." Aware of the possibility of "knavery" in the testimony before them, judges weigh the testimony of the two sides in the case before them. They probe the witnesses for contradictory statements to detect coherence. So far as they are able, they discover the character of the people testifying. When it is a question of a historical point of legal precedent, judges, according to Whiston,

> ...do not themselves pretend to judge of the Reality or Obligation of any Ancient Laws, or Acts of Parliament, from their own meer Guesses or Inclination, but from the Authenticness of the Records which contain them ... And owning that Ancient Laws, and Ancient Facts, are to be Known not by Guesses or Supposals, but by the Production of Ancient Records, and Original Evidence for their Reality.[121]

Whiston is supremely confident that the testimony of revelation will withstand the most skeptically minded judge (provided he is also fair-minded and not debased by nature). Whiston ends the *Astronomical*

Principles with a sermon on the "temper" of mind necessary for the interpretation of Scripture. So long as interpreters are objective, Whiston believes that their "shrewdest Abilities" and "utmost Sagacity" in weighing the testimony of Scripture will prove those testimonies to be "convincing Evidence." However, the comparison of these highly probable historical accounts of revealed religion with the certain truths of natural religion inferred on the basis of the "Divine Volume" of nature is the final test. Where the certain Newtonian system of nature and the certain inferences of natural religion conflict with the testimony of Scripture, then we must honestly acknowledge "a terrible Suspicion, that the latter is either false, or at least interpolated."[122] This is never the case in practice, even though it is theoretically possible. In practice, as Whiston shows in *A New Theory* and again in his *Astronomical Principles,* scriptural descriptions are internally coherent, based on the most ancient records, and perfectly compatible with the explanations of Newtonian science. Conjectures based on Scripture, which are necessarily rooted in the less trustworthy evidence of the human testimony contained in revelation, can never attain the same level of certainty as the inferences of science or natural religion. But when that testimony is corroborated by the explanations provided by science, as practice shows it always is, we have all the more adequately established reasons for giving our assent.

The *Astronomical Principles of Religion* forms a natural sequel to *A New Theory.* The former is ostensibly a cometary cosmology, the latter another exposition of the Newtonian design argument. But both display a much more basic concern with how to interweave revealed religion with the more sure and certain methodology of natural science and natural religion. Consequently, this interpretation points to the need for a broader concept of Newtonianism. Newtonianism includes, in Whiston's explicitly sanctioned extension of it, distinct arguments based on a specific method of biblical interpretation that are intended to prevent such "Ill tempered" deists as Blount and such good but confused and dangerous men as Burnet from severing the connection between the truths of natural religion and the myths and fables found in the testimony of Scripture.

Only by emphasizing the historical context of Whiston's dispute with the deists about Genesis is it possible to move toward a full understanding that Whiston believes that he is defending Scripture – not just explaining it away – and that this project is Newtonian in the sense that Newton agreed with Whiston's overall project as well as with his specific methodology. Whiston used Newton's method of interpreting Genesis, which Whiston may have arrived at independently fifteen years after Newton's correspondence with Burnet, to strike an interpretative balance between vulgar fideists for whom every word of the Bible was divinely derived (and hence, absolutely true) and the scoffing deists such as Blount who made jokes about biblical history and reduced it to the same status as a nursery rhyme or bedtime story. This was the project that Whiston understood himself to be embarked upon in his *New Theory*, which was, in all probability, "well approved" by Newton, and it would be anachronistic to claim that Whiston's approach is antiscriptural or a secularizing reinterpretation. It is what he claims it to be: a rational defense of Scripture in an age of transition in which he, as a Newtonian exegete, strove for the middle ground between vulgar theism and gross deism.

III. Newtonianism: inchoate or nonhomogeneous?

Michael Hunter, relying upon Kubrin's claim that John Keill is more truly representative than Whiston of Newton's views on hypothesizing about the divinely provident causes of phenomena described in the Bible, has recently attacked Margaret C. Jacob's thesis that Newtonianism was a self-conscious ideology with a clearly definable theological consensus; that it was readily identifiable as in alliance with the position of latitudinarian, Low Church moderates; and that it was, through the Boyle Lecture series, aimed at combating nasty Hobbist and "Epicurean" atheism, which, people felt, threatened to undermine the political and social order by destroying the authority of biblical revelation. As far as Hunter is concerned, the case of Keill, who attacked Whiston's *New Theory,* shows that "though there was undoubtedly a 'Newtonian' lobby in the earth sciences, it is an exaggeration to postulate any unanimity in such matters, since opinion remained inchoate."[123]

There is, however, a distinction between "inchoate" and "nonhomogeneous" that, the historical situation seems to suggest, must be made with regard to Keill, whose membership in the Newtonian inner circle is a fact.[124] Keill does indeed attack many of Whiston's scientific explanations of Moses' descriptions, but he acknowledges that Whiston "has made greater discoveries and proceeded on *more Philosophical Principles* than all the Theorists before him have done."[125] In fact, Keill accepts Whiston's hypothesis that a comet passed close by the earth around the time of the Deluge. He does not, however, believe that it was a sufficient natural cause to produce the effects described by Moses. Keill is most immediately concerned with showing flaws in Whiston's naturalistic explanations of Scripture history intended to demonstrate that such events are the direct and immediate work of divine special providence. Keill believes that Whiston goes too far in the direction of Burnet and that by contriving a naturalistic explanation of such events as the Deluge, "without the extraordinary concurrence of the Divine power,"[126] he has played into the hands of the very "atheists" he has tried to combat. Keill is, in general, a direct literalist who believes that biblical history is both a historically accurate description and, because literally true, a satisfactory scientific account as well, which reveals specially provident act after specially provident act, miracle after miracle.

Keill attacks Whiston where, in his hypothesizing, he moves too far away from Moses' literal words. The atmosphere of a comet, for example, is clear and luminous and does not, therefore, correspond to the "darkness without form and void" that was the chaos. Whiston argues that in cooling the cometary chaos had lost its luminescence and become dark. Keill catalogs many other difficulties in which Whiston draws away from the divine word to give a mechanistic explanation of it. Thus, even if the forty days of rain at the time of the Deluge really did issue from a comet's tail, Whiston's hypothetical explanation about where all this extra water went is inadequate. Whiston had hypothesized that it flowed back into the earth in the new fissures resulting from the cracking of the crust. Keill calculates the amount of water necessary and concludes that such a process would require 1,786.4 years, and hence,

> ...altho' Mr. Whiston has been pleas'd to ridicule my fondness for Miracles, yet since all the natural causes he has assign'd are so vastly disproportionate to the effects produc'd, he may at last perhaps be convinc'd that the easiest, safest and indeed only way is to ascribe 'em to Miracles.[127]

But just as Whiston does not reject all miracles,[128] neither does Keill reject all mechanistic explanations of phenomena. Whatever the source of floodwaters and by whatever means they were removed, Keill agrees with Whiston's hypothesis that fossils are the remains of once-living organisms destroyed by the Deluge. Both agree that Scripture is in some sense accurate. Whiston resorts to miraculous explanations only when there is no adequate mechanical hypothesis, whereas Keill refuses ever to depart from the literal words of Moses.

These differences do not make it impossible to recognize an orthodox Newtonian, as Hunter claims, nor do they demonstrate that Newtonianism was "inchoate." They do demonstrate that Keill, to the degree to which he departed from Newton's and Whiston's rule and "that which is clearly accountable in a natural way, is not without reason to be ascrib'd to a Miraculous Power" (Postulate II), was a Newtonian dissident. But Whiston's postulate, to which Newton subscribed, and Whiston's view that God's providence is best evidenced by an orderly progression of natural law, with which Newton also agreed, remain the criteria by which this facet of Newtonianism must be measured. Kubrin's conclusion that Keill's attack on Whiston's *New Theory* in 1698 is evidence that Keill played a "critical role in defining Newtonianism"[129] is unconvincing. In addition to the stark parallels between Whiston's method of interpretation and Newton's and their concurrence regarding the superiority of natural order for demonstrating both general and a kind of special providence, the best argument to the contrary is that in 1698, the year in which Keill's attack appeared, Newton intervened decisively in Whiston's career. He summoned Whiston from his parish at Lowestoft to be his substitute at Cambridge, gave him the "full profits" of his Lucasian Chair, helped him to become a candidate for that post when he relinquished the chair, and, by his recommendation of Whiston to the heads of colleges ensured that he would be his successor. There is consequently no reason to doubt Whiston's account in his *Memoirs* stating that when a manuscript of *A New Theory* was shown to Newton, "on whose principles it depended," Newton "well approved it." And, as I shall show in the next chapter, Newton continued to be active in Whiston's career by proposing the topic for Whiston's 1707 course of Boyle Lectures. But I shall also show that Hunter is correct when he claims that it is mistaken to speak in terms of the "triumph" of Newtonianism. Even before 1720, the deist threat to revealed religion – fueled, ironically, by the very success of the Newtonian design argument – was a present and growing danger.

3

WHISTON'S NEWTONIAN ARGUMENT FROM PROPHECY; DIVINE PROVIDENCE; AND THE CRITICISM OF ANTHONY COLLINS

Margaret C. Jacob has contended that the Boyle Lectures were the primary platform for the triumphant application of the design argument of Newtonian natural religion to the social needs of Restoration England. According to Jacob, Low Church moderates and Newtonian scientist-theologians (all of whom she labels simply "Newtonians") offered the stable Newtonian model of the physical universe as a model for civil and religious stability in a society that they believed had been "chosen by God to accomplish the Protestant Reformation."[1]

Jacob's thesis, with its emphasis on the Boyle Lectures as the medium through which Newtonian natural religion was integrated with Low Church, Whiggish social ideology, has drawn the criticism of Michael Hunter, who argues, in general, that Newtonianism was too inchoate to become the foundation of any social ideology whatever and, in particular, that the Boyle Lectures reflect the lack of any such orthodox Newtonian-latitudinarian application of the Newtonian design argument. Hunter justly observes that "by no means all the Boyle Lectures had any scientific component at all."[2] Hunter, following Jacob, identifies the only possible socially significant aspect of Newtonianism as the design argument (which infers providentially designed order in the physical universe). Then, because many of the Boyle Lectures do not refer to the Newtonian design argument or the natural philosophy that underlies it, Hunter concludes that Jacob's thesis is open to question.

I am not so much interested, in this chapter, in the social application of Newtonianism as I am in yet another aspect of what constitutes this overlooked feature of Newtonianism as a method of biblical interpretation, the underlying principles of which are shared by Newton and Whiston. In the preceding chapters we examined Whiston's attack on any form of allegorical interpretation of scriptural history, especially the Mosaic account of creation, which Whiston, following Newton, argues is literally true and in perfect accord with Newtonian scientific principles. Underlying Whiston's *New Theory* is his basic Newtonian principle of interpretation: "The Obvious or Literal Sense of Scripture is the True and Real one, where no evident reason can be given to the contrary."[3] We have noted, too, how Whiston's method of biblical interpretation is used finally to demonstrate the divinely provident nature of creation in Whiston's powerful statement of the Newtonian design argument in his *Astronomical Principles of Religion, Natural and Reveal'd.* I argued that it was legitimate to label Whiston's method of biblical interpretation "Newtonian" because it so closely resembled Newton's own sketchily stated method in his letters to Burnet and because Newton actively promoted Whiston's career after the method was published in the preface to *A New Theory.*

In this chapter, we will examine further evidence that Newtonianism encompassed a literal method of biblical interpretation. Whiston's 1707 Boyle Lectures are simply another instance of his application of this method. In these lectures, entitled *The Accomplishment of Scripture Prophecy,* Whiston tried to show the literal, singly determinate, nonallegorical, historical fulfillment of several biblical prophecies in the particular person of Jesus and to demonstrate in the process God's specially provident control and direction over his creation. Newton, I will argue, not only shared Whiston's literalistic method of biblical interpretation but also actively encouraged Whiston to apply it to the interpretation of historical prophecies regarding the messiah, from the platform of the Boyle Lectures. Whiston's Boyle Lectures do not indeed contain "any scientific component at all," as Hunter points out. Yet they must be considered an important aspect of the whole fabric of Newtonianism, since Newton, after securing Whiston's appointment to his own vacated Lucasian Chair, in all probability suggested to Whiston the very topic of these lectures.

Jacob, I believe, is correct when she argues that the purpose of disseminating this more widely focused social Newtonianism was to foster a stable social order by preserving society securely in its Christian moorings, safe from the taunts of scoffing deists. Ironically, the deists generally accepted the Newtonian design argument and its inference of a generally provident creator-deity, even while they harshly mocked the "fables" of Jesus. By widening the focus of Newtonianism beyond the scope of the design argument, I hope to supplement Jacob's thesis by showing that Whiston's Newtonian argument from prophecy, in addition to providing a timely, stabilizing model for the "world politic" drawn from the Newtonian model of the "world natural," also utilized the literal Newtonian method of biblical interpretation to demonstrate that the generally provident creator of the design argument is also the specially provident deity of Scripture, who is still a direct agent in the affairs of his creation.

Jacob overstates her case, however, when she argues that by 1720 the "Newtonian version of liberal Protestantism prevailed against the church's opposition."[4] Certainly, Jacob shows that Newtonian natural religion had an important influence on the latitudinarian vision of the social order in the first decades of the eighteenth century. However, the significance of Whiston's intellectual career lies in his attempt, sanctioned by Newton, to cobble natural religion together with revealed religion. The exegetical aspect of Newtonianism did not triumph, and Whiston's *Essay Towards Restoring the True Text of the Old Testament* (1722) provoked Anthony Collins's *Discourse on the Grounds and Reasons of the Christian Religion* (1724). Collins's book is the most famous deistic attack on Whiston's Newtonian method of interpreting biblical prophecies. The fire-storm of controversy that Collins provoked between 1724 and 1730 indicates that this aspect of Newtonianism did not share in the success of Newtonian natural religion.

In Section I of this chapter I trace the background to Whiston's Boyle Lectures. In Section II, I examine the contents of Whiston's Newtonian argument from prophecy and Anthony Collins's criticism of them. Section III serves as a brief conclusion.

I. The background to Whiston's argument from prophecy

As Newton's letters to Burnet suggest, Whiston's *New Theory* merely echoed Newton's original criticism of Burnet's "allegorical" reading of the Mosaic history of creation. From an early stage in his career, however, Newton was also interested in fabricating his own version of the design argument. Newton is entirely serious when he tells Bentley, concerning the publication of Bentley's Boyle Lectures (the first ever delivered) that "when I wrote my treatise about our Systeme I had an eye upon such Principles as might work with considering men for the beliefe of a Deity & nothing can rejoyce me more than to find it useful for that purpose."[5]

Bentley's Boyle Lectures utilized the complex Newtonian "Systeme" of natural laws, especially that of gravity, to infer the necessity of a generally provident, geometrically skilled creator to cause such order. Whiston testifies to the ready acceptance of Bentley's statement of the Newtonian design argument among the deists but points out that Bentley's lectures also produced an unexpected and undesirable side effect:

> ...certain Persons, not over-religiously dispos'd [were] soberly asked, after Dr. *Bentley's* remarkable sermons at Mr. *Boyle's* lectures, built upon Sir *Isaac Newton's* Discoveries, and level'd against the prevailing Atheism of the Age, *What they had to say in their own Vindication against the Evidence produc'd by Dr. Bentley?* The Answer was, *That truly they did not well know what to say against it, upon the Head of Atheism: But*

> *what,* say they, *is this, to the fable of Jesus Christ?* And in confirmation of this Account, it may, I believe, be justly observed, that the present gross *Deism,* or the Opposition that has of late so evidently and barefacedly appear'd against Holy scriptures, has taken its Date in some Measure from that Time.[6]

This comment goes to the heart of the mocking spirit of disbelief in Scripture shared by the new sect of freethinkers who abounded in London's coffeehouses. Whiston had much experience with deists. The *New Theory,* with its insistence that "the obvious or Literal Sense of Scripture is the True and Real one," is as much an answer to Charles Blount's posthumously published attack on the first chapters of Genesis (in *The Oracles of Reason,* 1693) as to Burnet's *Theory.* And in the final chapter of the aptly titled *Astronomical Principles of Religion* (1717), Whiston recapitulates the idea that "the overbearing Light of Sir *Isaac Newton's* wonderful Discoveries" in Physicks and Astronomy have afforded certain knowledge of the role of God in the natural world.[7] Whiston's discussion of these discoveries in the natural sphere – which is nothing more than the design argument – is addressed "to All," especially "if thou beest a Sceptick or Unbeliever, either as to Natural or Revealed Religion." The law of gravity, he argues, is the "Effect of a Supreme Being," and its certainty, based on such unquestionable evidence as astronomical observation and geometrical reasoning, has routed the skeptical natural systems of "*Democritus, Epicurus, Ptolemy, Tycho, Cartes,* Mr. *Hobbs,* or Spinoza."[8]

Whiston believed that deism had arisen because the discoveries of Newton had so effectively destroyed the systems of earlier scientific theories. Such earlier theories of the world

> ... now plainly appear, from certain Evidence, to be not only false, but absurd; contrary both to common Sense, and to the Known Laws and Observations of sound Philosophy; and ... he who will now be an *Atheist,* must neither understand the Principles either of Physicks or Astronomy.[9]

Because the general providence of God can no longer be doubted – that is, because the design argument is considered to be irrefutable – what *can* be doubted is revealed religion and its specially provident deity. Deists are theists, but they are not Christians. Of deism, which Whiston defines as "the Denial of the Scriptures, and of Divine Revelation," Whiston says that

> ... this modern infidelity is not properly owing to any new Discovery of the want of real Evidence for Reveal'd Religion, or of the Falsity of any of the Known Foundations of it; but to the like Necessity of Affairs, and the Impossibility of supporting the former, and worser Notions.[10]

Since the foundation of natural religion appeared to him to be utterly certain, Whiston attributed the deists' desire to press the attack on revealed religion to ignorance, perversity, or madness. He knew a great many of the deists personally, and his relations with them gives insight into his own character and the rise of the deist movement.

Before he was expelled from Cambridge, Whiston had befriended Thomas Woolston. Woolston is now remembered as the deist author who extended Collins's attack on allegorical prophecy to miracles.[11] Whiston remembered him "in his younger days" as "a clergyman of very good reputation, a scholar, and well esteem'd as a preacher . . . and beloved by all good men that knew him."[12] Whiston's *Memoirs* trace the progressive development of Woolston's deism and attributes its beginnings to the time when

> . . . he most unfortunately fell into *Origen's* allegorical works; and poring hard upon them, without communicating his studies to anybody, he became so fanciful in that matter, that he thought the allegorical way of interpretation of the scriptures of the *Old Testament,* had been unjustly neglected by the moderns.

When these views were rebuffed, "he grew really disorder'd [of mind] and . . . he was accordingly confined for a long time; after which, tho' his notions were esteem'd in part the effect of some such disorder, yet did he regain his liberty." In his pamphlets Woolston began to insinuate that Jesus' miracles were only allegorical, and this, together with his refusal to reside at his college, eventually cost him his Cambridge fellowship, even though Whiston interceded on his behalf. After Woolston was thrown out of the university, "The government fell upon him, and had him indicted in *Westminster Hall,* for blasphemy and profaneness." Whiston still stood by his friend and went to the attorney general's office to

> . . . give him an account of poor Mr. *Woolston,* and how he came into his allegorical notions; and told him, that their common lawyers would not know what such an allegorical cause could mean; offering to come myself into the court, and explain it to them, in case they proceeded: but still rather desiring they would not proceed any farther against him.

The matter probably would have been dropped, but Woolston then published another pamphlet explicitly ridiculing Jesus' miracles as mere allegories and denying their historicity in any literal sense. Though Whiston still felt that Woolston's "disordered mind" had driven him to such extravagances, this act was too much for him, and he no longer interceded publicly on Woolston's behalf. Finally, the court proceeded against Woolston for blasphemy, found him guilty, and sentenced him to a fine and imprisonment. Whiston concludes his anecdote of Woolston's career:

> In short, he seemed to me to have so confounded himself with his allegories, and so pleased himself when he found one gentleman, Mr. *Anthony Collins,* to affirm nearly as he did, tho' with a quite different design, that Jesus Christ dealt in allegorical prophecies, though not in allegorical miracles; that before he died he seems hardly to have known himself whether he really believed the christian religion or not.

Another of the lesser deists befriended by Whiston was Thomas Chubb, a philosophical tallow-chandler. Whiston again traces in his *Memoirs* the degeneration of one of his friends from a "judicious christian" into the "directly opposite character of one of the most foolish and injudicious of our modern unbelievers, as a comparison of his *first* and *last* books will demonstrate." Chubb's first book was *Eight Arguments on the Supremacy of the Father* (1714) and was edited by Whiston, who "took care of the correcting" of this work. "Without a learned education," Chubb was a man of natural philosophical ability, and his first book created a stir. On the strength of his performance in that work, considered singular because of his lack of formal training rather than for its contribution to the antitrinitarian controversy, Whiston secured for his untutored friend a job with Sir Joseph Jekyll. With his characteristic spleen, Leslie Stephen describes Chubb's new post as part that of servant and part that of literary plaything.[13] The period while he held this position, which he soon lost as his writings became more deistic, was the only time in Chubb's life during which he did not support himself by working at his trade of tallow-chandler. Chubb's book launched his literary career, but with each new pamphlet (there were to be more than fifty in all) his criticism of revealed religion grew more extreme. Whiston, hearing of this tendency, wrote Chubb predicting that he would turn into a "sceptick" if he continued in this path, "which his answer did by no means clear; and which his later writings too fully justify."

Of the more important deists, Whiston knew Anthony Collins particularly well because he had frequently dined with him, during 1711, at Lady Caverly's house in Soho Square, after his banishment from Cambridge. Lady Caverly was a deeply committed Christian who often invited to dinner divines such as Whiston or Samuel Clarke, while her skeptical paramour, Sir John Hubern, invited deists such as Collins or Matthew Tindal. This mixed company, according to Whiston, "used to meet, and to have frequent, but friendly debates, about the truth of the bible and christian religion."[14]

Whiston records similar skepticism concerning the Scriptures among some members of the clergy. Dr. Hare, for example, "was for laying wagers about the fulfilling of scripture-prophecies."[15] Dr. Cannon, according to Whiston, was a notorious unbeliever who once said "that if he were at *Paris,* he would declare himself a roman catholick; and if he were at *Constantinople,* he would declare himself a musselman, as taking religion to be an engine to promote peace in this world, rather than happiness in the next."[16]

These deists, within and especially without the ranks of the clergy, posed a particular philosophical problem as well as appearing to threaten the stability of the existing social order by ridiculing the revealed moral foundations on which it seemed to stand. The problem was that of balancing two sorts of divine providence: general providence and special providence. Most deists agreed that in the original act of creating the world out of nothing and setting up the laws of nature, God had exercised general providence. The heavens, they thought, do indeed reveal God's handiwork in his generally provident design of the world and nature at the time of creation. The design argument, with its many instances of order and regularity drawn from observations of nature, reveals this generally provident architect-creator. The deists accepted this argument in general, and with it the generally provident divine architect.

But what is God's role in his own carefully designed, generally provident creation after the initial act of creating and structuring the system? According to Christians, God continues to intervene in the fabric of his creation in acts of special providence. Christians maintain that the Bible is a literally true historical record of God's continuing, specially provident *control* of the world. The biblical testimony of miracles performed and prophecies fulfilled shows that God has not absconded but continues his specially provident direction and governance over both the "world natural" and the "world politic."

The deists accepted the design argument and its inference to a generally provident designer-architect. But they rejected biblical testimony that, to traditionalists, appeared to show a specially provident Lord of Creation who continues to oversee his fiefdom (he is aware, after all, of everything that happens, down to every fall of a sparrow) *and* to intervene in it directly through miracles and fulfilled prophecies. What, the deists ask, has the generally provident deity of the design argument to do with the "fables" of Jesus Christ or, indeed, the "fables" of Moses? The deist movement was a sustained attempt to undercut the evidence from the Bible of God's specially provident nature while cheerfully accepting that whoever built the universe long ago built geometrically and well. John Leland summarizes the antibiblical thrust of the deist movement: "No man that is not utterly unacquainted with the state of things among us can be ignorant, that in the present age, there have been many books published, the manifest design of which was, *to set aside revealed religion*."[17]

In Chapter 2, I traced Whiston's battles with the deists regarding the proper method of interpreting the Genesis creation story and his belief that the operation of natural law reveals a continuously active, interventionist, specially provident deity. But beyond this, Whiston also specifically wished to preserve God's power to break natural law through miracles and fulfilled historical prophecies (the latter of which are a kind of miracle, because they exceed the merely human predictive power of a prophet).

Although attacked by Keill for going too far in providing naturalistic explanations for miraculous events recounted in the Bible, Whiston did believe in God's ability to shatter the laws of nature. For example, he "vindicated" Phlegon's account of the earthquake at the time of Christ's passion as a genuine product of specially provident divine intervention – that is, a miracle.[18] He also took great pains to distinguish the "false" miracles of Simon Magus from the "true" miracles recounted in Scripture.[19] Whiston, finally, believed in the miracles of Jesus and believed that the Apostles continued to have specially provident power to contravene the laws of nature in such supernatural acts as the curing of disease by the laying on of hands.[20]

Although Whiston believed in specially provident miracles wrought by God in supernatural contravention of natural laws, he preferred to emphasize fulfilled prophecies as a means of demonstrating divine special providence and thereby countering the deist attack on this scripturally founded doctrine. The reason for this preference, as will be shown more fully in Chapter 5, was that with the new science the laws of nature were becoming so detailed and well defined that less room was left for miraculous events, whereas fulfilled prophecies seemed to demonstrate God's continuous agency in nature without – apparently – involving him in disrupting it. The meaning of God's past actions, which are described in the historical record of the Bible, becomes clear through his acts in the less distant past or in the present. The progressive revelation of divine history in fulfilled prophecies reveals a specially provident deity continuously active in his creation. The demonstration of fulfilled prophecies becomes the ideal weapon to use against the scoffing deists who accept the generally provident deity of the Newtonian design argument but ask what this argument (and this divine architect) have to do with the "fables" of Jesus.

As John Locke shaped the argument from prophecy just prior to the opening of the eighteenth century, the "reasonableness" of Christianity consists in the way Jesus fulfills prophetic predictions that the Messiah would "evidence his Mission" by working miracles. For Locke, miracles and prophecy stand in a mutually supporting synthesis, the latter predicting the former and the former confirming the latter. Of the time of Jesus, Locke states:

> The Spirit of Prophecy had now for many ages forsaken the Jews: And though their Commonwealth were not quite dissolved, but that they lived under their own Laws, yet they were under a Foreign Dominion, subject to the *Romans*. In this state their account of the time being up, they were in expectation of the *Messiah;* and of deliverance by him: Which gave them hopes of an extraordinary Man yet to come from God, who with an Extraordinary and Divine Power and Miracles, should evidence his Mission, and work their Deliverance. One great Prophet and worker of Miracles, and only One more, they expected; who was to be the Messiah.[21]

Throughout the course of the century a general tendency to separate miracles and prophecies and to consider each as the basis of an independent argument against the deists becomes evident. Unlike Locke, who appropriated miracles into his version of the argument from prophecy as one of the prophesied predictions fulfilled in Jesus, the other major intellect dominating the intellectual gateway to the eighteenth century, Isaac Newton, relied almost exclusively on fulfilled prophetic predictions to demonstrate the special providence of the God of Christianity. Newton's general opinion of miracles is that they are not really disturbances of the laws of nature but are simply unusual occurrences that excited wonder in the minds of ignorant ancients. "For miracles," states Newton, "are not so called because they are the works of God, but because they happen seldom and for that reason excite wonder."[22]

For Newton, the fulfillment of biblical prophecies is a more certain demonstration of the divine origin and authority of Scripture and of the specially provident nature of the Christian God than miracles. Of the prophet Daniel, Newton flatly states that "to reject his Prophecies, is to reject the Christian Religion. For this religion is founded upon his Prophecy concerning the Messiah."[23]

Newton considers the fulfilled prophecies contained in the historical record of the Bible as the primary foundation for the Christian religion. The problem arises immediately, however, that the greater part of these works are written in language that does not immediately refer to the historical world. Newton regards prophecy as a prediction but recognizes the difficulty of telling what exactly is predicted in the visions of Daniel and Saint John. What these prophets had predicted of the future had by the eighteenth century, Newton thought, almost all come to pass and is therefore a legitimate concern for a prophetic historian who seeks to interpret these visions correctly. Newton operates in the tradition of historians in the seventeenth century who interpreted biblical prophecy in this way. Chief among these was Joseph Mede, who clearly states that these prophecies must be interpreted within the framework of the continuous process of history. Daniel and Revelation are connected for Mede (and Newton), in that Daniel is the "apocalypsis contracta and the Apocalyps [i.e., the Book of Revelation] Daniel explicate."[24] The scope of these particular visions is nothing less than the panorama of history "from the beginning of the Captivity of Israel, until the Mystery of God should be finished."[25]

As an interpreter of prophecy, Newton accepts this view of the sweep and range of the historical prophecies in Daniel and Revelation and works out the language of these prophets, using Mede's basic principles that a day in prophetic language is equal to a year in current terms, and the idea that Revelation contains a kind of recapitulation of Daniel. As a guide to writing history, Newton works out a lengthy method that is basically a conversion table. In this table the symbols of the prophets are interpreted to demonstrate how specific historical events, including the rise and fall of postbiblical kingdoms and especially the first arrival of the Messiah, are precisely predicted in the Book of Daniel.

In his literal hermeneutic method, Newton assumes what Frank Manuel has called a "vital point of contact"[26] between the images contained in biblical prophecies and historical events in sociopolitical history. For Newton, an event mentioned in a fulfilled historical prophecy in Daniel as transpiring in the "world natural" refers simply to a future historical event in the "world politic." Thus, says Newton, "ascending towards heaven, and descending to the earth, are put for rising and falling in power and honour; ... descending into the lower parts of the earth, for descending to a very low and unhappy estate; ... great earthquakes, and the shaking of heaven and earth, and passing away of an old one, for the rise and ruin of the body politic signified thereby."[27] This basic method, in conjunction with the long-hallowed dating principle of interpreting a day as a year in prophetic language,[28] enables Newton to write history, or "chronology," as he prefers to call it. With this method, Newton spells out how the prophecies in Daniel and Revelation are fulfilled in the history of the biblical kingdoms and the postbiblical kingdoms:

> The whole scene of sacred prophecy is composed of three principal parts: the regions beyond *Euphrates,* represented by the first two beasts of *Daniel,* the empire of the *Greeks* on this side of the *Euphrates,* represented by the Leopard and by the He-Goat; and the empire of the *Latins* on this side of *Greece* represented by the beast with ten horns.[29]

Of more immediate concern to the specialized argument from prophecy regarding the Messiah, Newton also uses this method to show how the predictions made in Daniel's prophecy of 70 weeks (Dan. 9:24–5) to the coming of the Messiah are exactly fulfilled in Jesus. The entire religion of Christianity, Newton states "is founded upon his [Daniel's] Prophecy concerning the Messiah."[30] Given that the 70 weeks in Dan. 9:24–5 means 70 weeks of years, or 490 years, Newton explains how Jesus fulfills Daniel's prediction. His method of calculating this makes clear why chronology was of such great importance to him:

> Now the dispersed *Jews* became a people and city when they first returned into a polity or body politick; and this was the seventh year of *Artaxerxes Longimanus,* when *Ezra* returned with a body of Jews from captivity, and revived the *Jewish* worship; and by the King's commission created Magistrates in all the land to judge and govern the people according to the laws of God and the King, *Ezra* vii. 25. There were but two returns from captivity. *Zerubbabel's* and *Ezra's;* in *Zerubbabel's* they had only commission to build the Temple, in *Ezra's* they first became a polity or city by a government of their own. Now the years of this *Artaxerxes* began about two or three months after the summer solstice, and his seventh year fell in with the third year of the eightieth *Olympiad;* and the latter part thereof, wherein Ezra went up to *Jerusalem,* was in the year of the *Julian Period* 4257. Count the time from thence to the death of Christ, and you will find it just 490 years.[31]

Because the 70 weeks must be "marked out for your people and your holy city" (Dan. 9:24), the beginning of the 490 years must be dated from the time of Ezra, because only with Ezra do the returned exiles form a polity. From then until the death of Christ was 490 years, leading Newton to believe that Jesus fulfilled the prediction of Daniel's visionary prophecy and thus was the promised Messiah. For Newton, as for Locke, to prove this true is to prove Christianity true.

In passing, I must refer to Newton's views about prophetic predictions in the Bible that had remained unfulfilled even down to his day, especially the apocalyptic predictions in Revelation. At this point, it is enough to mention that regarding predictions yet to be fulfilled, Newton again follows Mede in exercising great caution about attempting to interpret such prophecies *before* their completion in history. Of the coming millennium, Mede says, "But here (if anywhere) the known shipwrecks of those who have been too venturous should make us most wary and careful, that we admit nothing into our imaginations which may cross or impeach any Catholick Tenet of the Christian Faith."[32] The church hierarchy had, since the time of Augustine, discouraged attempts to date the coming revolutions predicted in Revelation, and Newton, likewise, is generally very cautious about interpreting the exact date of the fulfillment of an apocalyptic prediction. Newton insists that the "time is not yet come" for understanding the predictions of biblical prophecy that have not yet come to pass, such as the "main revolution" predicted in Revelation. The following passage shows Newton's caution in this regard, while expressing the belief that we have a sufficient guarantee of "God's providence" in the historical and messianic prophecies that *can* be interpreted:

> There is already so much of the Prophecy fulfilled, that as many will take pains in this study, may see sufficient instances of God's providence: but then the signal revolutions predicted by all the holy Prophets, will at once both turn men's eyes upon considering the predictions, and plainly interpret them. Till then we must content ourselves with interpreting what hath been already fulfilled.[33]

Newton's *Observations upon the Prophecies,* in its final form, was not published until 1733, five years after his death. Whiston informs us that the draft versions, which are much more coherent and much longer than the edited version that finally was printed, were composed while Newton was "a young Man, or not more than 40, or at the most 50 years of Age."[34]

Whiston's Boyle Lectures and his many other works on the historical completion of biblical prophecies apply in detail his master's briefly stated, posthumously published method of interpretation. Whiston's lectures, delivered in 1707 and entitled *The Accomplishment of Scripture Prophecy,* aim to complement the Newtonian design argument and illustrate to the deists precisely how the "fables" of Jesus are related to it. Whiston not only utilizes Newton's hermeneutic method in this endeavor, but he very likely had the help of Newton himself, as a consideration of Newton's role in the process of selecting Boyle lecturers shows.

According to very strong circumstantial evidence put forward by Henry Guerlac and Margaret C. Jacob, Newton played a behind-the-scenes role in determining the course and character of Bentley's inaugural Boyle Lectures.[35] First, from a text of David Gregory's, Guerlac and Jacob establish that Newton was already thinking of some form of public dissemination of the results of his natural design theology for the strengthening of natural religion *before* his famous correspondence with Bentley. In a fragment of conversation that he reported from December 1691, Gregory says,

> In Mr. Newton's opinion a good design of a publick speech (and which may serve well at one Act) may, be to shew that the most simple laws of

> nature are observed in the structure of a great part of the Universe, that the philosophy ought then to begin, and that Cosmical Qualities are as much easier as they are more Universall than particular ones, and the general contrivance simpler than that of Animals plants etc.[36]

A "publick speech" about the efficacy of his "cosmical" discoveries for revealing providential design in the created universe was therefore a concern of Newton's when, in the next month (January 1692), he went down to London for Boyle's funeral.

Next, from a letter of Samuel Pepys's, then president of the Royal Society, Guerlac and Jacob establish that Newton was invited to dine with John Evelyn, a leader among the group charged, in a codicil to Boyle's will, with the responsibility for establishing a series of public lectures "for proving the Christian religion against notorious Infidels." Pepys, in a letter dated January 9, 1692, invited Evelyn to dinner with other members of the Royal Society to discuss the course of science in England now that Boyle was gone: "Pray lett Dr. Gale, Mr. Newton and my selfe have the favour of your company to day, forasmuch as (Mr. Boyle being gone) we shall want your helpe in thinking of a man in England fitt to be sett up after him for our Peiriskius besides Mr. Evelin."[37]

Because of the historical facts established by these two texts – (1) that Newton considered a "public speech" about his version of the design argument; (2) that he then met a month later with an executor of Boyle's will charged with instituting a public lecture series whose purpose was to strengthen the Christian religion; and (3) that Bentley the first Boyle lecturer, with the help of correspondence from Newton, actually used the lecture series as a platform for stating Newton's design argument – Guerlac and Jacob conclude that Newton probably played a behind-the-scenes part "in the selection of Richard Bentley as the first lecturer, and encouraged – if he did not suggest – the theme of Bentley's last sermons."[38]

But Newton was as concerned with preserving divine revelation as he was with establishing his own scheme of natural design theology. Thus he criticized Burnet for allegorizing the Mosaic history of creation and helped to advance Whiston's academic career on the basis of Whiston's parallel attack on Burnet. Newton also, as we have seen, was strongly interested in the argument that Christianity is founded upon the literal fulfillment of messianic prophecy.

Newton's interest in the literal argument from prophecy, and the tremendous furor surrounding the critical deist attack on such prophecy interpretation at this time, suggest that Newton may have attempted to play a behind-the-scenes role in Whiston's Boyle Lectures against allegorical prophecy interpretation. In support of this hypothesis, we have Whiston's own statement that his Boyle Lectures and his subsequent works on prophecy interpretation were first suggested to him by Newton himself. Whiston traces his interest in the interpretation of fulfilled messianic prophecies and his particular method of interpretation directly to Newton. Whiston states that his particular version of the argument from prophecy was not

> the Product of my own Thoughts originally; but was intimated first to me, when I was Young, and was in the common Opinion, by a very Great Man, who had very exactly studied the Sacred Writings; who, among some other valuable Hints he than gave me, that have been of great Use to me in my Studies since that Time, made this very just Observation, "That the Christian Commentators had hurt the Christian Religion, by ascribing the several Prophecies of the Old Testament to Persons living under that Dispensation; which properly and truly belong'd to none but to the Messiah himself." Which Observation was to me, so far as I can now remember, the first proper Occasion of my Enquiries of this Nature, and of the most satisfactory Consequences of those Enquiries afterward.[39]

Whiston published these words in 1725, when Newton was still alive; presumably he did not identify Newton by name to avoid giving pain to his former patron by associating him with himself, now a notorious and controversial heretic. But in his *Memoirs,* long after Newton's death, Whiston reveals that this "very Great Man" who stimulated his interest in the literal interpretation of prophecy was in fact Sir Isaac Newton. Whiston recounts that the "sceptick," Dr. Hare, "once blabb'd out" that he "feared Christ and his apostles were so weak, as to depend on the double sense of prophecies for the truth of christianity," until Whiston, "upon Sir *Isaac Newton's* original suggestion, shewed [Hare, Bishop Chandler, and Dr. Clarke] the contrary."[40]

The discovery that Whiston's 1707 Boyle Lectures on *The Accomplishment of Scripture Prophecy* and his subsequent works on that topic were originally suggested to him by Newton himself is highly suggestive when correlated with Guerlac's and Jacob's evidence of Newton's probable role in engineering Bentley's 1692 lectures concerning the Newtonian design argument. If Newton was not actually involved in the selection of Whiston as the Boyle lecturer in 1707 – and while there is no direct evidence that he was, it does not seem unlikely on the face of it, given Newton's later boosting of Whiston's career – it seems very likely that he suggested to Whiston the topic of his series of lectures.

It seems entirely plausible, on the basis of this data, to view Newton as a genuine *eminence grise* who utilized the Boyle Lecture series to put forward by proxy his own theories not only regarding the Newtonian design argument but also regarding the Newtonian method of interpreting messianic prophecies in the Bible. Whiston's lectures on the progressive revelation of the Messiah were intended to confound the scoffing deists and reveal a specially provident deity still active in governing his creation, thus balancing the generally provident architect-creator inferred in the design argument.

Concerning the "design of God" in providing mankind with so many examples of fulfilled prophetic predictions in Scripture, Newton says in his book on prophecies that "the events of things predicted many ages before, will then be a convincing argument that the world is governed by providence."[41] Whiston, in his lectures, echoes Newton and agrees that the fulfilled prophecies signify "*the design of God's providence,* which was the reinstating the ruin's [ruinous] affairs of faln Men, and the destruction of that wicked but potent Empire which the Devil had set up, by the coming of the promis'd Messias, and the gradual advancement of His Kingdom."[42]

II. Whiston's Newtonian argument from prophecy; biblical criticism; and Anthony Collins's reply

By the eighteenth century the definition of prophecy had been codified by Samuel Johnson in his famous *Dictionary*. According to Johnson, the noun *prophecy* means "a declaration of something to come; prediction." The verb *to prophesy* means "to predict; to foretell; to prognosticate."[43] Nevertheless, it is not clear what counts as the fulfillment or completion of such a prediction. A theory arose with the early Christian Fathers regarding the differing ways in which prophetic predictions may be considered to be fulfilled. The third-century Christian Platonists of Alexandria, especially Clement and Origen, wishing to dig beneath the surface of the literal scriptural narrative to reveal the timeless philosophical meaning of texts, used an allegorical approach. The presence of a deeper, hidden level of meaning in Holy Writ led Origen to a large-scale use of allegorical interpretation to show how the Old Testament messianic prophecies typify or prefigure Christianity. Always, to Origen, the simple literal fulfillment of prophecy in the historical world is less important than the deeper spiritual level of meaning. Thus the Exodus from Egypt may be seen on the literal level as a real (and directly providential) divine action to free the nation of Israel from bondage. On the allegorical level, however, the Exodus typifies or symbolizes Christian baptism, which is a fuller and more spiritually significant act of divine rescue.[44]

Origen's and Clement's allegorization of scriptural prophecy tended to trivialize the literal, historical level of meaning, but they always acknowledged this level. Augustine insists on both as equally important, however, and this view was eventually codified by Thomas Aquinas, who asserts that the many multiple meanings of a scriptural text are ultimately founded on the literal level.[45]

Whiston claims that Grotius was the first of the "moderns" to reduce this multiple level of meanings into a "regular system" with regard to messianic prophecies with his interpretation of Isa. 7:14 and Matt. 1:22–3.[46] In Isaiah, Ahaz had prophesied that a son named Immanuel would be born to the House of David from a virgin. Saint Matthew claimed that the Gospel story of the Virgin Mary's giving birth fulfilled Ahaz's prophecy. Grotius, who thought that the literal, historical fulfillment of Isaiah's prophecy had occurred in Isaiah's own time with the birth of Immanuel, claimed that this prophecy could apply to Jesus only allegorically or typically.

Whiston is outraged by the notion that biblical prophecies must be interpreted in terms of an elaborate system of multiple meanings. In the same way that Burnet's allegorizing of Moses' historical narrative trivializes it and lays it open to skeptical doubts so, too, Whiston thought, the allegorization of messianic prophecies trivializes them and converts them into mere "fables" of no interest to modern, deistic freethinkers. Whiston's basic principle of interpretation in his *New Theory,* however, as we have seen, was that "the Obvious or Literal Sense of Scripture is the True and Real one, where no evident Reason can be given to the contrary."[47] This literal approach to the historical narrative of Moses is followed by Whiston in all of his works on the interpretation of messianic prophecy, and it is this literal, single-meaning view of scriptural prophecy that constitutes Newton's "original suggestion" for Whiston's Boyle Lectures. Even though the language of the prophets is "peculiar and enigmatical," by following Mede's and Newton's careful hermeneutic principles (e.g., a day equals a year) one can, they believed, interpret the literal, historical fulfillment of biblical prophecy without recourse to any hidden, allegorical meaning. Whiston comments, "I observe that the Stile and Language of the Prophets, as it is often peculiar and enigmatical, so it is always single and determinate, and not capable of those double Intentions, and typical Interpretations, which most of our late Christian Expositors are so full of upon all Occasions."[48]

For Whiston, almost all the messianic prophecies "have been properly and literally, without any recourse to Typical, Foreign and Mystical Expositions, fulfill'd in Jesus."[49] As far as Whiston is concerned, the concept of multiple levels of meaning is a fanciful hypothesis akin to Aristotelian entelechies and Cartesian vortexes, and such careless hypothesizing must be abandoned by the Newtonian biblical exegete, just as analogous "vain hypotheses" have been abandoned (so it is claimed) by the Newtonian natural philosopher. Whiston argues that

> till the learned Christians imitate the learned Philosophers and Astronomers of the present Age; who have almost entirely left off Hypotheses and Metaphysicks, for Experiments and Mathematicks; I mean till they be content to take all Things, that naturally depend thereon, from real Facts, and original Records; without the Byass of Hypothesis, or Party, or Inclination; 'till then I say I verily believe that Disputes and Doubts, Scepticism and Infidelity will increase upon us.[50]

Whiston's various writings on already completed historical prophecies describe nearly three hundred prophetic predictions as having been clearly and directly fulfilled.[51] This aspect of Whiston's Newtonian biblical exegesis met with a warm reception. For his yeoman work against the free-thinkers' ridicule of the fables of Jesus, Whiston states that Bishop Gilbert Burnet even forgave him his Arian heresy.[52] Whiston claims that his literal, single-sense interpretation of messianic prophecy "intirely recovered the late lord *Abercorn,* a considerable member of the Royal Society, from his scepticism or infidelity; as he fully owned to me himself long before his death."[53] Samuel Parker, in a long, enthusiastic review of Whiston's Boyle Lectures, states that Whiston is justified "in the Charge he brings against those expositors who make the Prophecies of the Old Testament to relate but in a secondary sense to the *messias* and the times of his kingdom."[54] Even the arch-deist Anthony Collins, with whom Whiston often dined and argued religion, agrees that to suppose that a single passage has only one level of meaning "is to proceed by the common rules of grammar and logic."[55]

But there is another facet of Whiston's literal Newtonian exegesis of scriptural prophecy that does not fare so well. While Whiston believes that *most* messianic Old Testament prophecies, as they are preserved in our present copies of the Bible, can be quickly shown to apply in one, and only one, literal sense to Jesus, he is aware that there are some "ancient Knots and difficulties" with interpreting a small number of the prophecies in this mode. He believes that problems arise in literally interpreting these particularly difficult prophecies because of corruptions and mistakes in our copies of the Old Testament. He is therefore immediately embroiled by his Newtonian method of prophecy interpretation in attempting to recover the true text of the Bible, which he attempts to do in his *Essay Towards Restoring the True Text of the Old Testament And For Vindicating the Citations made thence in the New Testament* (1722). It was this work and this aspect of Whiston's Newtonian biblical interpretation that provoked

Collins's explicit rejection of the argument from prophecy and his pointed criticism of Whiston's method, in *A Discourse on the Grounds and Reasons of the Christian Religion* (1724).

Whiston is convinced that "the present text of the Old Testament is, generally speaking, both in the History, the Laws, the Prophecies . . . or as to the main Tenor and Current of the Whole, the very same now that it ever has been from the utmost Antiquity."[56] His lists of more than three hundred literally fulfilled prophecies is proof of the general accuracy of the present text. A very few prophetic predictions concerning the Messiah, however, seem to require an allegorical level of meaning if they are to be applied to Jesus. The prophecy in Isa. 7:10–16 that a son named Immanuel would be born to the House of David is precisely such a case, as Grotius points out. Because St. Matthew claims that Jesus himself was the fulfillment of this prophecy, such modern expositors as the "great *Grotius* . . . are forced to go roundly and frequently into that strange Notion of the *double* sense of Prophecies."[57] The birth of Immanuel seven hundred years before Christ seems to have been the literal, historical fulfillment of this prophecy.

But for Whiston, to suppose a double meaning is to make the written word contrary to common sense and the rules of grammar. If prophecies are allowed more than one level of meaning, Whiston says, "We can never be satisfy'd but they may have as many as any Visionary pleases," or, as William Nicholls puts this same point, "If we should once allow this typical or allegorical way of explaining Scripture, one might as well prove the history of Guy of Warwick out of the first chapters of Genesis."[58] Unlike the greatest Catholic biblical critic, Richard Simon, who took refuge in the authority of the Church to decide the issue when such knotty problems of exegesis arose,[59] the Protestant Whiston was left with no choice but to try to locate the source of the error by the light of common sense.

Whiston's commonsense method for restoring the true text is to rely on the oldest and, in his view, the necessarily least corrupt versions of these works. All modern versions of the Old Testament (including the Hebrew text used as the primary source today for translations into critical editions and vulgar-language copies) derive from the second century A.D., when Jewish religious leaders compiled a text from such manuscripts as had survived Titus Vespasian's destruction of the Temple in A.D. 70. This text is called the traditional, or Masoretic, text – from the Hebrew word for tradition, *massorah*. Whiston blames these copyists for purposely introducing corruptions into the Masoretic text, because so many of the apparent mutations "prove to be in Texts, wherein the Christian Religion and the ancient Christian quotations are concerned." That the Greek Septuagint version has also been corrupted Whiston demonstrates by giving examples of the considerable "variety, both the Words and Sense, of a vast number of Texts in the several Copies of our Septuagint."[60] Resolving that the true text will cut through these "Ancient Knots" of interpretation that have so long been a stumbling block to Jews and men of reason, he devotes the rest of the *Essay* to outlining a plan to follow to achieve this purpose. In the case of Matthew's citation of Isaiah, for example, Whiston believes that he can rely on the quotations of this part of Isaiah as it appears in the *Apostolical Constitutions*. In this work the text of the prophecy is rearranged in a way that makes the birth of a child from a young woman in Ahaz's day appear to be an event entirely separate from the messianic prophecy (made to the House of David, not to Ahaz) that a virgin of that family would conceive and bear a son called Immanuel. Whiston therefore asserts that

> . . . the Reasons why I venture to say the *Jews* seem to have corrupted or dislocated their Copies here, are, because Coherence and Context do now look much disorder'd: because it is certain, that in the Copies of *Justin Martyr*, and of *Tertullian* in the Second and Third Centuries, these prophecies were not wholly in the same Order wherein they now stand; and because the *Apostolical Constitutions*, in the first century, quote two of them much after the same manner as they are here set down, and not as they lye in the present Copies.[61]

Collins charges "that a *Bible restored*, according to Mr. W.'s Theory, will be a mere WHISTONIAN BIBLE, a BIBLE confounding and not containing *the True Text* of the Old Testament."[62] This is true if one regards Bible texts, and especially messianic prophecy, as mere fable and hence unreconstructable in principle. But Whiston's method of using the most ancient and hence *uncorrupted* versions of a text as the basis for the official text is the most obvious and fundamental principle of textual criticism. Modern authorities now agree that the *Apostolical Constitutions* is a fourth-century forgery, but it is only one of several other texts that Whiston regards as "helps" provided by providence in the task of restoring the true text, and these other texts are still regarded today as useful for this purpose.

Apart from the *Apostolical Constitutions*, which Whiston does date incorrectly, he proposes utilizing several other sources when the official version of the Old Testament is finally prepared. For example, Whiston states that "the several *Greek* Editions and MSS. of the *Septuagint Version* now extant, with other *Translations* made anciently from it"[63] will be of great help in recovering the original text if used with care. So, too, Godfrey R. Driver, the director of the committee responsible for preparing the modern English translation known as *The New English Bible*, says of the Septuagint that "it is valuable for the recovery of the original Hebrew, because it is based on an underlying Hebrew text older than the Masoretic, and it often preserves the correct reading in passages where our Hebrew manuscripts are manifestly in error."[64] Whiston recommends consulting the Samaritan Pentateuch. So, too, does Driver. Whiston suggests having a look at the "few Remains we have of the later *Greek Versions;* particularly those of Aquila, Symmachus, and Theodotian, anciently preserv'd in the *Hexapla* of Origen." Driver points to these texts as well.[65]

Finally, in an innovative recommendation that reveals Whiston's brilliance as well as his commonsense scientific approach, he proposes that a "Great Search should be made in all Parts of the World for *Hebrew* Copies, that have never come into the Hands of the *Masoretes;* and for Greek Copies of the . . . vulgar *Septuagint* version, read in the Churches all the First Ages of Christianity; or any Parts of Them: For if any Books of these Kinds can be recovered, they will probably, so far, be uncorrupt Copies of the Old Testament."[66] This exhortation is itself almost prophetic because of the recent recovery of the Qumran scrolls, which contain, for one thing, a complete text of Isaiah in Hebrew dating from the second century B.C.[67]

In fact, with the exception of the *Apostolical Constitutions,* whose illegitimacy has only been proved since Whiston's time, the most recent full-scale translation of the "true text" of the Old Testament has followed Whiston's method almost exactly.[68] *The New English Bible* may be fairly considered, by virtue of the method utilized to "go behind" the present Masoretic text in order to derive the "true text" for purposes of translation, as a genuine "Whistonian Bible." Of course, if the method is the same, the results are very different from those that Whiston hoped for. The Qumran version of Isaiah, for example, is almost identical to the Masoretic text, leaving unresolved the question of in what sense Jesus fulfills the prediction made in Isa. 7:10–16 cited by Saint Matthew. Driver says that "in spite of this wealth of ancient versions, and even when the earliest known form of the text has been established, many obscurities still remain in the Hebrew Scriptures."[69]

But Whiston, a true son of the Newtonian Enlightenment, is confident that scientific common sense will succeed in restoring the earliest text of the book by carefully weighing the different versions, and that this restored text will establish the literal argument from prophecy to its rightful eminence as a rationalistic bulwark to revealed religion. He concludes the work on this note: "As the *Miracles* of our Blessed Saviour, and his Holy Apostles, have been all along *one* great and undeniable Argument for the Truth of the Christian Religion; so will that *other* Argument taken from the *fulfilling of the ancient Prophecies of the Old Testament,* belonging to the

Messias, in Jesus of Nazareth, deserve hereafter to be esteem'd equally great and undeniable."[70]

Whiston considers the *Essay* as the keystone in his protracted argument for Newtonian literal interpretation of prophecy. With the method for the recovery of the true text clearly delineated, Whiston is certain that it is only a question of time before the skeptical cavils of the deists and the wrongheaded and irrational allegorical readings of his opponents in the clergy will be dismissed at a stroke. In this book, then, Whiston believed he had erected the latest and most advanced "New Machine of War" against the skeptics and blindly fideistic clergy that he had first announced in his *New Theory of the Earth*.

The deist Anthony Collins is quick to rebut Whiston's literal Newtonian interpretation of prophecy by homing in on Whiston's enterprise of restoring the true text of the Old Testament. Collins charges that Whiston's "Warmth of Temper" disposes him "to receive any sudden Thoughts, any Thing that strikes his Imagination, when favourable to his preconceived Scheme of Things, or to any new Scheme of Things, that serve, in his Opinion, a religious Purpose."[71] This disposition leads Whiston to the arbitrary reconstruction of the Old Testament, which Collins labels a "Whistonian Bible" – a Bible that, on the basis of older variants, eliminates any need for an allegorical level of meaning in prophecy by revising it so that prophecies apply to Jesus literally.

Collins, with most other thinkers of the first half of the eighteenth century, agrees that the core of Christianity is that Jesus is the Messiah. He also perceives that the only "proof" worthy of the name that can establish this core tenet is the argument from prophecy. If this argument is valid, "Christianity is invincibly establish'd on its true Foundation." Conversely, if the argument is invalid, "then is Christianity false."[72] And if Whiston believes he can show the argument to be valid by his textual criticism, which supposedly demonstrates the literal fulfillment of Old Testament messianic predictions in the person of Jesus, then Collins believes he himself can show the argument invalid by casting doubt on any attempt, literal or allegorical, to connect the Old and New Testaments. Exploding Whiston's "theory" of textual criticism is one of the main themes of Collins's *Discourse*.

Accordingly, Collins attacks the sources Whiston proposes to utilize as "helps" in the derivation of the original text. In his brief dismissals of such sources as the Syriac Old Testament, Collins is heavily dependent on the scholarship of Richard Simon, who says, for example, that the Syriac version is "less exact than the Hebrew Text of the Jews, and the Greek Version of the Septuagint."[73] But, as we have seen, Whiston recommends several sources. Concerning Whiston's suggestion that a search be instituted for the most ancient and therefore least corrupt versions, Collins offers the following truism: Such works, says Collins, "no where appearing, and being themselves *unrecover'd,* cannot, till *recover'd,* be of any Use *towards restoring a true Text.* They are themselves to be *restored,* in order to restore a true Text."[74] Collins observes that after examining all of the prophecies cited by the Apostles as having been fulfilled in Jesus he has not discovered a single one that is capable of the literal and direct mode of interpretation. He consequently explicitly rejects Whiston's mode of literal interpretation:

> ... prophecies cited from the Old Testament by the authors of the New, do so plainly relate, in their obvious and primary sense, to other matters than those which they prove, in that sense, what they are produc'd to prove, is to give up the cause of Christianity to the Jews and other enemies thereof.[75]

For example, the writer of the Gospel of Matthew declares (Matt. 1:22) that Jesus fulfills the prediction of the prophet Isaiah (Isa. 7:14) that the virgin will conceive and bear a son. Unfortunately, as Collins points out, this particular prediction made to Ahaz had already been fulfilled in the birth of Immanuel seven hundred years prior to the birth of Jesus. And, anyway, inquires Collins, "How could a virgin's conception and bearing a son seven hundred years afterward be a sign to Ahaz[?]"[76]

Therefore, unless we are to give up the cause of Christianity, we must seek some method of showing how Jesus fulfills messianic prophecies not literally but typically or allegorically. The Gospel writers, Collins believes, would not have claimed that Jesus fulfilled a prophecy already fulfilled. To "explain" how we may hope to understand the rules for interpreting prophecies allegorically, Collins appeals, tongue in cheek, to the "learned Surenhusius," a Dutch writer who claims to have recovered the long-lost method of prophecy interpretation common among Jewish biblical scholars. Collins ridicules Surenhusius's great rediscovery of this method of deriving the "hidden" typical meaning from a biblical prophecy. Surenhusius's rules of interpreting prophetic language include such cryptographic embellishments as changing the entire word order; changing word order and adding other words; and "changing the order of words, adding words, and retrenching words, which is a method often used by Paul."[77]

Collins's point is that to show how prophetic predictions have been fulfilled typically or allegorically in Jesus we must resort to absurd rules of interpretation. And yet, because key messianic prophecies that Jesus is claimed to have fulfilled clearly refer, in our present, valid Bibles, in a primary and directly literal sense to some previous historical set of circumstances, we have no other choice.

Collins sets his argument up as a sort of dilemma of Hobson's choice. Either take the only path open to us in prophecy interpretation, that of Surenhusius, or forfeit prophecy. But in taking that path, prophecy, as an argument bolstering Christianity, is forfeit in any case because, according to Surenhusius's outlandish method, prophecies are made meaningless as they are rearranged to suit the requirements of the particular interpreter. One path leads to a "Whistonian Bible," the other to a "Surenhusian Bible."

The fact that Collins chose to attack Christianity by attacking the argument from prophecy indicates that he (like Locke, Newton, and Whiston) believed Christianity to be chiefly founded on this argument. The avalanche of protest shows how sensitive religious writers were to this line of attack. By Collins's count,[78] there were thirty-five replies to his *Discourse,* although three of these, all by Thomas Woolston, were supportive of Collins's argument. The other thirty-two were ferociously hostile and included Whiston's own reply, *The Literal Accomplishment of Scripture Prophecies.*

Regarding the prophecy of the "virgin" birth in Isaiah, Whiston does not so much argue against Collins as he does simply repeat his own position that the prophecy refers "properly and singly" to Jesus, citing as evidence again the variant reading in the *Apostolical Constitutions.* In this book, which is subtitled "Being a Full Answer to a late *Discourse*" (which it manifestly is not), Whiston resorts to his favorite technique of listing in one column the Old Testament prophetic prediction and in an adjacent column its New Testament "completion," as if relying on the sheer weight of the number of predictions and completions he has gleaned from the Scripture to carry the day for the argument from prophecy.

Leslie Stephen calls this "reply" of Whiston's to Collins's *Discourse* "singularly absurd."[79] And it is fair to point out that Whiston's entire argument from prophecy does seem to beg the question. The deists assume that any inference, made on the basis of a prophecy, that leads to the conclusion that Jesus is the Messiah is simply invalid, because such predictions are fables. Presenting more examples of such fulfilled predictions will not appear to the deists to provide a rational foundation for revealed religion – and yet this is precisely how Whiston answers Collins's attack on the argument from prophecy. Why, if Whiston realized that this was the deist position, did he attempt to answer Collins by simply repeating over and over his numerous examples of how Jesus "completes" the prophetic predictions on the basis of the "ancient facts and testimonies" of the Bible?

Instead of disdainfully dismissing Whiston's reply as "absurd," it would be better to try to piece together, relying on the hints he provides in his diffuse and unsystematic presentation, what Whiston believes he is accomplishing by replying in this way. When examined in this fashion, the question becomes not one of logic but attitude. I will not defend Whiston's logic but instead will explicate the blinkered outlook that prevents him from realizing that giving more examples of completed predictions does nothing to change their fabulous status in the eyes of the deists.

Whiston provides some hints of his attitude toward the proper method of reasoning against the deists in a later book growing out of the Collins controversy and entitled *A Supplement to the Literal Accomplishment of Scripture Prophecies* (1725). If Isaac Newton's formulation of his method and epistemological views is "truncated,"[80] Whiston's is positively cryptic, especially in his writings concerning the degree of certainty that his method of scriptural interpretation yields. But what clues he does provide are extremely suggestive. In the following text, Whiston declares that feigned hypotheses have no place in scriptural exegesis, where, as in the physical sciences, the experimental method must reign supreme; he also describes the sort of evidence that experiments of this kind require and suggests an analogy between what he is doing in interpreting prophecies and what experimental scientists of the Royal Society do in the physical sciences or what judges in the law courts do when hearing a case. Whiston states:

> Nor do I find that Mankind are usually influenc'd to change their Opinions by any Thing so much, as by Matters of Fact and Experiment; either appealing to their own Senses now; or by the faithful Histories of such Facts and Experiments that appealed to the Senses of former Ages. And if once the Learned come to be as wise in Religious Matters, as they are now generally become in those that are Philosophical and Medical, and Judicial; if they will imitate the Royal Society, the College of Physicians, or the Judges in Courts of Justice; (which last I take to be the most satisfactory Determiners of Right and Wrong, the most impartial and successful *Judges of Controversy* now in the World:) If they will lay no other Preliminaries down but our natural Notions, or the concurrent Sentiments of sober Persons in all Ages and Countries; which we justly call the Law or *Religion of Nature* . . . And if they will then proceed in their Enquiries about Reveal'd Religion, by real Evidence and Ancient Records, I verily believe, and that upon much Examination and Experience of my own, that the Variety of Opinions about those Matters now in the World, will gradually diminish; the Objections against the Bible will greatly wear off; and genuine Christianity, without either *Priestcraft* or *Laycraft*, will more and more take Place among Mankind.[81]

Whiston clearly believes that he is extending the Newtonian or Royal Society empirical method into the realm of revealed religion in his many books that purport to show how Jesus fulfills Old Testament predictions. For Whiston, as well as for many of his contemporaries, prophecies are predictions. Such prophetic predictions may be empirically verified as fulfilled or not fulfilled by real evidence that, in Whiston's case, is sought by searching the most ancient historical records, those that are closest in time to the event. In the field of interpreting messianic prophecy, such a procedure is an experiment, because ultimately these testimonies are based on the sense experience of the reporters. This explains Whiston's enthusiasm for such "primitive" documents as the *Apostolical Constitutions* and accounts for his proposal, in 1712, to set up a "primitive library" at his house in Cross Street in Hatton Garden. The closer the testimony to the event, Whiston believes, the more credible the evidence.

Armed with this method, Whiston attempts to thread his way between the Scylla of the double-talking members of the "Priestcraft" (whose allegorical mystifications of Scripture cause Whiston to wonder why they simply have not entirely denied "that there were ancient uncontested Predictions concerning the coming" of the Messiah) and the Charybdis of the "Laycraft" as represented by deists such as Collins. Because revealed religion has not so far been approached in scientific fashion, Whiston says, "Tis no great wonder that Christendom is so divided in their Sentiments about divine Matters, and that every Sect and Party finds their own peculiar dogmata there."[82]

Collins's charge that Whiston creates a "Whistonian Bible" is accurate. Whiston's belief that the text preserving the literal sense of prophecy will be the oldest is false, as the Qumran scrolls have shown; at least his hopes for this approach have not so far been justified. Furthermore, Collins scores heavily in his ridicule of Whiston's thesis that the Jewish copyists deliberately distorted their versions of the Old Testament. Nevertheless, as the reaction to Collins's book showed – by Collins's own count, his book provoked thirty-two hostile replies – the argument from prophecy remained stronger than ever and can, in the first half of the eighteenth century, be legitimately characterized as the "citadel of orthodoxy."[83] And, as I have shown, Whiston's method for reconstructing the true text of ancient Scripture on the basis of the most ancient texts may not resolve all difficulties, as he expected it to do, but it is the only method available and the one practiced by modern translators. In addition, many of Whiston's contemporaries, including Bishop Burnet and Samuel Parker, applauded Whiston's effort to maintain the literal version of the argument from prophecy. Most important, the whole enterprise had probably been suggested to Whiston by Newton. The reason why this aspect of Newtonianism, as practiced by Whiston, ultimately fell into disrepute is that Whiston applied it to the Arian heresy. To trace the social application of Whiston's Newtonian method of biblical interpretation is the theme of Chapter 4, where the results Whiston obtains by its use in political theory and in

Arian and millennial theology are examined. Here it is enough to note that critics soon noticed that the man who did not shrink from telling the deists how to reconstruct the true text of the Old Testament did not shrink from using the same method to tell his Anglican brethren how to reform their corrupt dogmas, such as that concerning the Trinity. Samuel Parker, who had favorably reviewed Whiston's Boyle Lectures in 1708, writes in 1709, when Whiston's Arian-inclined *Sermons and Essays upon Several Subjects* came out, that his historicist approach to revising the biblical canon to support his Arian views is "such a *Carybdis* . . . as is enough to frighten the vulgar sort of Christians and Scholars upon the *Scylla* of Popish Infallibility." Parker asks,

> To what purpose do you dispute with Mr. *Whiston* about what is in Scripture, when he runs out such a further length of Scepticism, as to make it a doubt what is Scripture? how many, not only Parts, but BOOKS, of the Old Canon are quite lost and extinct? and how many are yet to be added to those still extant and Receiv'd? . . . And can any body of men be sure they are Christians, that know not whether they believe all the Original Articles of the Christian Faith? So that, upon the whole, Mr. W has in a few lines (and to the same purpose is meditating perhaps Volumes) endeavour'd to establish such notions and opinions about the Word of God, as make it highly improbable, that for several hundred years past there has been, or could be, a Church of *Christ* upon Earth. I hope these Pestilent Positions are as some people have thought, only the workings of a distemper'd brain.[84]

III. Conclusion

We have seen in this chapter how Whiston attempts, with Newton's blessing, to combat deism through the literalistic and providentialist argument from biblical prophecy. He is in a solidly orthodox stance when he disputes with freethinkers such as Collins or radical millenarians such as the French Prophets. Two of the most important French Prophets, John Lacy and Francis Mission, had read Whiston's Boyle Lectures with approval, even though Whiston had disparaged their activities in his sermons. In 1713, Whiston met with Lacy and several other of the prophets and attempted to show them that their visions and impulses were perhaps supernatural but were nevertheless evil in inspiration. Besides being unable to interpret any difficult prophecies, such "wild agitations are rather signs of daemonical possessions, than of prophetick status."[85] In these endeavors, Whiston was moving in the mainstream of contemporary scholarship; but as soon as he applied his Newtonian method of biblical interpretation to the burning political and social issue of church authority or the politically sensitive religious issues of Arianism and millennialism, he was labeled an "enthusiast," as we will see in detail in the next chapter.

It seems clear that Margaret C. Jacob is correct in her attempt to widen the concept of Newtonianism beyond natural science. The previous two chapters have been devoted to supplementing this aspect of her thesis by widening Newtonianism still further to include a literalistic method of interpreting the Bible, whether it be Moses' account of creation or messianic prophecies. Second, Jacob is correct in her emphasis on the importance of the Boyle Lectures in this process. The fact that the topic of Whiston's Boyle Lectures was *The Accomplishment of Scripture Prophecy,* rather than the design argument, does not mean that these lectures had nothing to do with Newtonianism. The topic was originally suggested by Newton, and it is likely that it was suggested to balance Bentley's generally provident creator-architect deity with the specially provident Lord God of the Bible.

Hunter's criticism of Jacob's claim for a Newtonian-latitudinarian triumph after 1720 is justified, however. In the next chapter we will see how Whiston's self-martyrdom on the point of his Arian heresy was not a triumph, since his offensive was crushed by a combination of High Church Tories and latitudinarian Low Church bishops. Furthermore, the deist pressure against this aspect of Newtonianism did not disappear with Collins. Chapter 5 will clarify Newton's and Whiston's positions with respect to the deist movement and, ultimately, trace the fruition of the deistic attempt to prise apart general and special providence, natural and revealed religion, in the work of David Hume.

4
WHISTON'S NEWTONIAN BIBLICAL INTERPRETATION AND THE RAGE OF PARTY, RADICAL ARIANISM, AND MILLENNIAL EXPECTATIONS

To this point we have traced Whiston's Newtonian attitude toward the relationship between natural philosophy and religion without analyzing in any detail the religious and political context in which these attitudes developed. Having established Whiston's basic intellectual stance, we are now in a good position to evaluate Margaret C. Jacob's ambitious interpretative perspective that relates the religious implications of Newtonian natural philosophy to the political and religious context of Restoration England. Even prior to the events of 1688–9, natural philosophers and divines of the Royal Society such as John Wilkins and Robert Boyle had emphasized the providential role of God in first ordering and then supervising the course of nature.[1] But in the context of the renewed political instability engendered as a consequence of the Glorious Revolution, Jacob argues that the Newtonian popularizers – brilliant young scientist-theologians such as Richard Bentley, Samuel Clarke, and William Whiston – championed the religious implications of specifically Newtonian natural philosophy and produced a "latitudinarian," Low Church, liberal, and rationalistic version of Anglicanism that appealed to the newly emerging class of monied city men of commerce by convincing them that the providentially ordering and guiding God of nature, demonstrated by the Newtonian design argument, was on their side. For Jacob, "latitudinarian" does not refer so much to the seventeenth-century "men of latitude" as to the "Newtonians" of the first two decades of the eighteenth century who forged the links between the religious implications of the Newtonian design argument and the monied commercial interests advanced by the Whig Party. With their Newtonian scientific theology, these Newtonian latitudinarians provided the monied men of trade and commerce, and the Whig politicians who served their political interests, with a model of the universe that was, above all, well designed and stable. Jacob summarizes the relationship between the latitudinarian Newtonians who are the heroes of her saga and the Whig ideology in the following statements of her central thesis:

> The ordered, providentially guided, mathematically regulated universe of Newton gave a model for a stable and prosperous polity ruled by the self-interest of men. That was what Newton's universe meant to his friends and popularizers: it allowed them to imagine that nature was on their side.
>
> . . . . .
>
> The latitudinarians grafted the New Philosophy on to their social ideology, integrating both into English thought at precisely the time when modern and capitalistic forms of economic life and social life were gaining ascendancy.[2]

Jacob's description of the mutually supportive linkage between Newtonian design theology and the Whiggish market society, though its presentation is open to criticism, seems compelling. More problematic is her identification of all aspects of Newtonianism with latitudinarianism and her ambitious extension of this thesis to her ultimate conclusion that by 1720, the latitudinarian Newtonians had utterly triumphed over both their scientific opposition ("Epicurean" atomists and Cartesians) and the opposing social and political ideologies of enthusiastic freethinkers and high-flying Anglican Tories.[3]

Whether Jacob is correct in this aspect of her thesis depends finally on what people of the time meant by Newtonianism. Jacob is correct to point out that Newtonianism included a sociopolitical component. But as I have argued, Newtonianism also included a method of biblical criticism that definitely had important political and social ramifications, as well as extremely heterodox religious consequences. The political and social ramifications of this version of Newtonianism survived and served, as Jacob argues, to underpin the restored political and social order, but the heterodox religious theories that also flowed legitimately out of Newtonian biblical criticism – especially Arianism and millennialism – did not. A sanitized version of Newtonianism survived, but it was changed so radically in the ideological conflicts with the freethinkers and the High Church Tories that to speak of a "triumph" is an exaggeration that simplifies what was in fact a very complex process in which the opposition to the Newtonianism of the first decades of the eighteenth century exacted a high penalty from the Newtonian popularizers who starred in Jacob's epic. Not least, Isaac Newton himself was cowed into silence by the kind of social pressure and economic penance exacted from Clarke and, especially, Whiston.

In Section I of this chapter, I seek to amplify Jacob's thesis about the importance of the sociopolitical implications of Newtonian natural philosophy by illustrating Whiston's application of Newtonian biblical interpretation to Whig political theory. It is a fair criticism of Jacob's presentation that despite her emphasis on the Newtonian contribution to establishing a stable and firm social order, her explanation of precisely how this is accomplished, beyond the analogy with the stable order implicit in Newton's God-designed system, is vague. Geoffrey Holmes notes that "to find discussion of those social changes, tensions and anxieties of post-revolution England, which are presented [in Jacob's book] as the inescapable context of the 'Newtonian' intellectual contribution, being carried on almost exclusively in generalities is especially frustrating."[4] Once again, an analysis of Whiston's work in the area of political theory helps to clarify the nature of the linkage between emerging Whig political theory and Newtonian biblical interpretation as understood by Newton and developed and applied by Whiston to the fields of earth history, design theology, the argument from prophecy, and, finally, to a rationale for the justness of dethroning one king and replacing him with another. In marked contrast to the Tories and High Churchmen who believed that a king's rightful authority was hereditary in nature, Whiston utilized his Newtonian method of biblical interpretation to show that a king's right to rule is a providential grant from God that is bestowed contractually by the "Choice and Recognition of the People."[5] The contractual *and* providential nature of the divine right of kings is the topic of Whiston's *Scripture Politicks* (1717). Whiston adapted his Newtonian method of biblical interpretation to the service of primarily Whig political doctrine.

Whiston had many personal contacts with Whig thinkers. He was befriended after his ouster from Cambridge and given economic support by such Whig notables as Sir Joseph Jekyll, Lieutenant-General James Stanhope, Sir Peter King, Nicholas Lechmere, Joseph Addison, and Sir Richard Steele.[6] This first section is intended to show, at least in part, why the Whig political theorists found Whiston's Newtonian biblical interpretation so congenial to their political viewpoint. Without doubt, as Holmes points out, the "Whig-Tory struggle in politics was by far the dominant fact of life for the post-Revolution Church."[7] The Newtonian method of biblical interpretation, as explicitly tailored to support moderate Whig political theory by Whiston, is at the heart of the Newtonian contribution to this struggle.

Whiston's work for the Whig cause was rewarded with money and offers of preferment. Sir Joseph Jekyll, the Whig Master of the Rolls and long-time benefactor of Whiston, went so far as to sound Whiston out about the possibility of Whiston and Clarke becoming bishops, "to amend what was amiss in the church." When Whiston pointed out that as an antitrinitarian he would not be able to subscribe to the Thirty-nine Articles, Jekyll replied that it would not be necessary for a bishop to subscribe.[8] Whiston refused this offer, predictably, on principle.

In Section II of this chapter, I trace the development of Whiston's Arianism in the context of the crucial Whig–Tory struggle within the church in the first decades of the eighteenth century. In his recent monumental study of Isaac Newton, Richard S. Westfall fleetingly discerned a "glimpse" of "one of the most advanced circles of free thought in England grouped around Newton,"[9] whose members shared the Arian view that Jesus was neither consubstantial nor coeternal with God the Father. This dimly perceived circle around Newton included Samuel Clarke, Richard Bentley, Edmond Halley, and Hopton Haines but featured as its most vocal spokesman William Whiston. Whiston's Arianism, like his theory of earth history and his Whig propagandizing, derived from his rationalistic, Newtonian method of biblical interpretation. After tracing the roots of this antitrinitarian revival in Whiston's Newtonian method of biblical interpretation, I will trace the reaction of the high-flying Tories within the Church to this heterodox challenge. I will argue that the attempt on the part of High Church Tories such as Francis Atterbury to try Whiston for heresy before the Convocation of 1711, which occurred just after the upsurge of high-flying Tory enthusiasm precipitated by the Sacheverell trial and a subsequent Tory sweep in the election of 1710, represented a genuine attempt by conservative clerics to silence and suppress an integral element of Newtonianism – that is, Arianism. As a supplement to the recent assertion by Richard G. Olson that High Church Tories were antagonistic to the excessive pride in human reason that formed the underpinning of the new science but that "identifiably Newtonian science escaped the most intense odium,"[10] I will contend that the High Church Tories saw Whiston's Newtonian biblical criticism as leading inevitably to Whiggish rejection of hereditary divine right, on the one hand, and to radical, antitrinitarian heterodoxy on the other. Whiston refers to himself in his *Memoirs* as a new "Luther."[11] Jonathan Swift's own skepticism about the ability of human reason to achieve knowledge in science, politics, or religion can be likened to Atterbury's persecution of Whiston. In the Tory newspaper established by Swift in 1711, the *Examiner,* Whiston and his Whig sponsors are criticized repeatedly for heterodox theological views. Whiston is said "to lead a party, as *Arius* did of old."[12] Further, "between Mr. *Whist-n* and *Christianity,*" the Tory writer charges, a cause is pending that the Whigs hope will be "decided in favour of the former."[13] Because this High Church, Tory criticism was leveled against Whiston's confident extension of Newtonian biblical interpretation to the revolt against the established Church's doctrine of the Trinity, the criticism is explicitly anti-Newtonian in character. In fact, Whiston paradigmatically typifies the overt linkage between pride in the achievements of Newtonian science and Whig political attitudes, as well as the radical religious heterodoxy that was so distasteful to Jonathan Swift. Swift's antiscientism was of a piece with Atterbury's persecution of Whiston; both were anti-Newtonian. Swift, not incidentally, personally detested Whiston; and he and the other Scriblerian Tories such as Arbuthnot took every opportunity to ridicule Whiston and his "projects."[14] Furthermore, as the Newtonian Arians were silenced one by one, the High Church counterattack to this form of Newtonianism achieved great success. Newtonian biblical criticism and its Arian and millennial consequences gradually became stigmatized as "enthusiastically" heretical in the eyes not just of High Church Tories but also of many latitudinarian bishops and serious laymen. The success of the High Church counterattack in ostracizing from the realm of serious thought this integral aspect of Newtonianism accounts, in my view, for Newton's growing distance from the Arian junta that he had helped to form. As Whiston and Clarke were being publicly pilloried, Newton remained silent, convinced, perhaps, that the wise would understand the rationality of antitrinitarianism without the necessity of jeopardizing his own political and social status as his followers had done.

In Section III of this chapter, I trace the development of millennialism out of Whiston's method of biblical interpretation. Newtonian physical science does reveal a generally stable natural order guided by a wise and providential deity, a deity who nevertheless can also cause catastrophic changes in the natural or political systems either through acts of general providence built into the fabric of the system at the time of creation or through specially provident, miraculous dispensations. Whiston's Newtonian ideology thus contained within it a potential for cataclysmic change that led, in the political realm, to the argument that the people have the right to choose their ruler when the ruler has exceeded his or her power and broken the contract to rule, and, in the theological arena, to the view that scientific reason – not received church doctrine – reveals the true nature of both God the Father and Jesus Christ. The radical consequences of Whiston's revolutionary method of interpreting Scripture is nowhere more evident than in his millennial expectations and in his view of his own role, as a new Luther, in bringing about a universal awareness of the imminence of the millennium as his Newtonian method of exegesis revealed it to him. Once again, millennialism is part and parcel of the Newtonian method of biblical exegesis. Whiston explicitly connected his compulsion to lead a revival of "primitive Christianity" – that is, Arianism – with both his Luther-like call for a "thorough reformation" of the church and the subsequent "hastening" of "the coming of our Saviour's Kingdom of peace and holiness."[15]

For Whiston, providential and contractarian political theory, reformed theology that culminates in Arianism, and the expectation of an imminent Second Coming were all connected by his Newtonian method of biblical interpretation. Contractarian political theory, shorn of its biblical foundation and thereby secularized, certainly "triumphed" in the eighteenth century, but the Arian and millennial aspects of Newtonianism dropped out of their originally Newtonian matrix in the course of the conflict between High and Low churchmen in the first decades of the eighteenth century.

In Section IV, I return, by way of a concluding summary, to my critique of Jacob's thesis about the identity of latitudinarianism and Newtonianism and its ultimate "triumph" around 1720.

I. Whiston's Newtonian biblical criticism and the rage of party

A large number of important studies concerning the shifting political tides after the Restoration, as Whigs and Tories battled each other in a wide variety of arenas, has appeared in the last few years. None has been as important as the work of Jacob in assessing the Newtonian contribution to political thought. Nevertheless, even Jacob is somewhat vague about the nature of the Newtonian intellectual contribution to the ideological battle within the ranks of the Anglican clergy and to the rise of Whig ideology. Her view is that the Newtonian version of a providential designer and his well-ordered creation provided a model for a providential and well-ordered society. Because of Newton's own reticence and desire to avoid controversy, it is difficult to go much beyond this generalization in assessing the impact of Newton himself upon developing Whig thought in the ferment of post-1689 politics. The truth of Jacob's assertion that Newtonian scientific theism served the emerging Whig party as a model for stability is best illustrated in the allegorical poem by the Newtonian scientist and Grand Master of the Grand Lodge of London Freemasons, J. T. Desaguliers, who, in his *Newtonian System of the World, the Best Model of Government* (1728), pleaded for stability and harmony upon the accession of George II. As soon as one considers that Newtonianism includes a particular historicist method of biblical interpretation, however, new light is shed on Newtonianism in the political context of the first decades of the eighteenth century.

The rage of party

The impact of recent scholarship has changed the way we regard the political aftermath of the Glorious Revolution.[16] Prior to this body of recent work, the standard interpretation of these events was that the Catholic King James II attempted to impose upon the English nation, on the basis of his hereditary divine right to rule, the Catholic religion. His opponents *immediately* determined that in this attempt he had violated his contract with the people and with Parliament, the true basis of his royal authority, and legally resisted him as a consequence. The people and Parliament deposed the Stuart tyrant in favor of a new contract with King William and Queen Mary, Queen Anne, and ultimately, the house of Hanover. Finally, after the fall of a succession of coalition governments, and of Harley and St. John's Tory government in 1714, a succession of Whig administrations, including Sir Robert Walpole's twenty-one-year premiership, achieved an extraordinarily high degree of political stability through their unchallenged political and ideological supremacy. On this view, the Whig stewardship was only occasionally disturbed by such anomalies as the Tory administration of 1710–14 and the Jacobite rebellions of 1715 and 1745, whose radical Catholic leaders wanted a Catholic king, and by small groups of Tories who still believed in James II's legal claim of divine hereditary right, in the political corollary of indefeasible succession, and in the absolute authority of the established church to define and enforce church doctrine. More recently, scholars have rejected this view, arguing that the whole train of events is far more complicated, especially with regard to the debates on these points within the Church itself.[17]

By the first decade of the eighteenth century, amid the heavy taxes levied to finance William's continental wars, the majority of Anglican clergymen felt that the Church had been betrayed by the bench of largely Low Church bishops who had too easily allowed the Whig politicians to push through liberal social policies detrimental to the Church of England. As far as these High Church Tory parsons were concerned, the aftermath of the constitutional crisis of 1688 had dramatically wounded the established Church. The effect of the Act of Toleration of 1689, for example, was to dramatically reduce church attendance, which led, in turn, to a decline in the authority of the parish priest over the morals and religious practice of his parishioners. Another effect of the Toleration Act deplored by the Tory clergy was the new freedom to attend worship services in the meeting-houses of Dissenters, which led to the proliferation of dissenting opponents to Anglicanism. The lapse of the statute requiring official censorship of books in 1695 increased the bitterness of Anglican vicars, who now saw scurrilous deist tracts ridiculing orthodox religious doctrines and, even worse, the orthodox clergy, published with impunity. But, at the root of the crisis within the clergy and within society at large was the grave question of conscience posed by the deposition of James II and the accession of William and Mary. Since the Reformation, Protestant churchmen had taught that within a king's realm, no earthly power was superior to the king because he possessed his right to rule from God. Hence the Church of England had always taught, as the underlying doctrines of social and political order, the principle of nonresistance to providentially constituted power (or "passive obedience"), and the principle of hereditary divine right with its corollary of indefeasible succession.

Twenty years after the Settlement of 1689, however, the question of the legality and morality of the Glorious Revolution was still not settled. Whiston's erstwhile Arian disciple, Benjamin Hoadley, voices the disquiet of both clergymen and laymen when he writes that

> the case of the Revolution is a public, national case of conscience . . . [and] it hath, I think, become absolutely necessary that, one way or other, it should be put as much out of doubt as possible. For the matter is now reduced to this; whether we lie under a national guilt, or not.[18]

Despite the best efforts of the latitudinarian bishops who replaced the nonjuring bishops, the nation was still riven into factions pitting High Church clergymen against Low Church bishops, Whig against Tory, and town against country. The latitudinarian bishops, such as Thomas Secker, William Sherlock, William Lloyd, Simon Patrick, and John Moore, worked diligently to provide a comprehensive rationale and defense of the Glorious Revolution,[19] but they met with great resistance from High Church Tories and the country squirearchy, who shared the Tory view that the Low churchmen in league with the Whig oligarchy had brought the nation to the brink of moral and social collapse. Geoffrey Holmes rightly states that the church reflected these polarized political and social coordinates:

> The division of its members into high and low parties, embroiling the clergy . . . at all levels and sucking them disastrously into the maelstrom of the Whig–Tory struggle in politics, was by far the dominant fact of life for the post-Revolution Church. From the 1690's onward it was all-pervasive and by 1705 overwhelming.[20]

When Dr. Henry Sacheverell preached an emotionally charged, highly inflammatory sermon in St. Paul's Cathedral on Gunpowder Day, 1709, much of the accumulated resentment of High Church Tory parsons and country squires was voiced. In his sermon "In Perils among False Brethren," Sacheverell propounds the conservative doctrines of unlimited hereditary divine right and passively obedient loyalty, while disavowing belief in the legality of toleration for religious Dissenters, in spite of the Act of Toleration of 1689. He rants against nonconformists, unlicensed schools, and Dissenters but reserves his most vitriolic language for those Whigs who argue that a king's right to rule rests on the consent of the people and that in extremely grave breeches of the social contract by the king (as the Whigs argued was the case with James II) the people might legally resist their ruler. As far as the actual Revolution of 1688 was concerned, Sacheverell accepts William's explanation that his accompanying army was only for his personal protection and not to assist in the resistance movement to James II, who had in effect abdicated. He thus explicitly accepts the fact of the Revolution, but his argument leads inevitably to the conclusion that the present rulers are usurpers and the present government illegitimate destroyers of the fabric of church and state.

Under pressure from the Court Whig and lord treasurer, Lord Godolphin (who later contributed money to Whiston), the Whig cabinet decided to impeach Sacheverell, for "high crimes and misdemeanours," in the High Court of Parliament. The Whigs apparently desired to put their interpretation of the constitution before the nation in a kind of national referendum with their High Church Tory opponents.[21] Whig theorists and statesmen such as Lord John Somers, Benjamin Hoadley, Sir Peter King, Nicholas Lechmere, and Lieutenant-General James Stanhope used the occasion of the high-flying Sacheverell's trial and the attendant war of pamphlets to make the Whig case. All of these influential Whig spokesmen were connected with Whiston.

Lord John Somers's sister married the eventual Whig Master of the Rolls, Sir Joseph Jekyll, who was Whiston's principal benefactor and who, as we have seen, once offered Whiston and Clarke bishoprics. Somers disagreed with Godolphin's decision to prosecute Sacheverell but nevertheless wrote against Sacheverell in which became a kind of Whig manifesto and vigorously stated the Whig position. Somers, who had helped in securing for Newton the position at the Mint and who was a good friend of Locke (who had served as Somers's adviser on the constitution), argues that history has revealed limited government to be most in accord with natural law. Those states and polities have longest survived and most flourished in which all power and authority has proceeded from the people, who also have the right to alter the succession, "upon very urgent causes." Utilizing profane history, Somers makes mincemeat of the doctrine of passive obedience by citing "above fifty Kings and nine emperors deprived of their Evil Government" in Europe alone.[22]

Whiston knew Benjamin Hoadley before Hoadley's activities as a Whig apologist finally gained him a bishopric. Hoadley felt in 1711 that the *Apostolical Constitutions* ought to be received in the church as the basis for church doctrine. Hoadley attended meetings in 1715 of Whiston's Society for Promoting Primitive Christianity but, to Whiston's disgust, renounced his Arian leanings when he became bishop of Bangor in 1716.[23] In 1710, as a pamphleteer for the Whig side in the Sacheverell affair, Hoadley argued for the divinely provident institution of a limited monarchy subject to the will of the people, who are the absolute judges "in *Case* of any doubt or *Competition,* to whom the *Right,* and *Title,* to the *Supreme* Government belongs."[24]

Sir Peter King (later Lord King) was one of the Whig managers at Sacheverell's trial. King later disappointed Whiston by forsaking the Arian doctrine of his youth, stated in King's *Enquiry into the Constitution, Discipline, Unity and Worship of the Primitive Church that flourished within the first three hundred years after Christ* (London, 1691), a book that had profoundly impressed King's nephew, John Locke. At the time of Whiston's heresy trial in 1713, King served as one of his defense counsels without fee.[25] At Sacheverell's trial, King confined himself to attacking Sacheverell's assertion that the Act of Toleration of 1695 was unreasonable and unwarrantable. King argued that the act did not lead inevitably to malicious heterodoxy or treason and that it was "one of the principal Consequences of the Revolution" and an essential element of the English Constitution.[26] Ultimately, King was appointed lord chancellor and totally lost his youthful zeal for primitive Christianity, prompting Whiston to remark, "What good christians will not be horribly affrighted at the desperate hazard they must run, if they venture into the temptations of a court hereafter?"[27]

Another of the Whig trial managers, Nicholas Lechmere, came down strongly in favor of a limited monarchy, arguing that this had been a part of the constitution since the Magna Carta.[28] Lechmere later aided King in defending Whiston at his heresy trial without fee.

Finally, one of Whiston's closest Whig supporters, Lieutenant-General James Stanhope, argued against Sacheverell at the trial, quoting Grotius and Hooker to show that the basis of the constitution was a compact between prince and people, which, in extremis, the people have a right to void.[29]

Whiston's Whig (and, in the case of Hoadley and King, Arian-inclined) friends succeeded by the narrow margin of seventeen votes in impeaching Sacheverell for "high crimes and misdemeanours." The Whig success was much diluted, however, by the derisively light sentence imposed by the Parliament. Sacheverell was suspended from preaching for three years.

Sacheverell's defense, managed by the high-flying arch-Tory Francis Atterbury, had been brilliant: Sacheverell had argued that he accepted the legitimacy of the Revolution and that he had only preached the traditional Anglican doctrine of passive obedience to the now established order and urged the need for the clergy to guard against heresy. Hence, he claimed, he could not understand his persecution by the Whig ministry. Sacheverell's trial provoked several ugly riots by Tory mobs. When he was given his minimal sentence, he also received a new living in Shropshire. On his journey there, he was met in the Midlands market towns along the way by adoring crowds stirred by his High Church Tory appeal of "Church in Danger." This enthusiasm was felt in Cambridge, where Richard Laughton, a senior proctor and Whiston's close friend, attempted to eject a group of men drinking at the Rose Tavern with the university's Tory M.P.'s. Laughton met with failure and was ridiculed by toasts to "Dr. Sacheverell."[30] In the election of the summer of 1710, the Tories rode this crest of popular enthusiasm to a sweeping victory. The Whigs had put their liberal constitutional theory before the public in the Sacheverell trial, and the public had rejected it, with uncomfortable consequences for vociferous public Dissenters such as Whiston.

In the context of this enormous social and political upheaval, it is misleading to state that the principal Newtonian contribution to the Whig side of this ideological struggle was confined to a well-designed, providentially ordered model of the universe encapsulated in the Newtonian design argument. Similarly, it is simply too vague to observe that Newton's effect on those Whig theorists lies in his stimulation to "greater freedom of inquiry."[31] Nor is it enough to state that the connection between this upheaval in the church and Newtonianism lies in the fact that the Whigs of the Sacheverell affair were all friends of Newton or Whiston.

The Newtonian contribution to the Whig–Tory, Low Church–High Church conflict was more specific. Utilizing his Newtonian method of biblical interpretation, Whiston became an apologist for the Whig principles of limited monarchy and religious tolerance.

Whiston's "Whig" apologetics

Whiston's "Whiggery" was typically idiosyncratic and reveals the danger in overemphasizing such labels. He was not a leveling or radically republican Whig, like John Toland or Anthony Collins, whose materialistic deism and radical politics he despised and actively combated.[32] Nor was he a Court Whig dedicated to pandering to the queen with ideological support of the Protestant monarchy in order to secure preferment. Whiston and Newton shared an abhorrence of the politicized nature of the corrupt modern church, as the following lengthy but significant anecdote attests. Whiston records the following "early" conversation (probably between 1692 and 1696) with Newton, which perhaps reveals why Newton at last himself set forth his version of the design argument in the "General Scholium":

> ...I early asked him, why he did not at first draw such Consequences from his Principles, as Dr. *Bentley* soon did in his excellent Sermons at Mr. *Boyle's Lectures;* and as I soon did in my *New Theory;* and more largely afterward in my *Astronomical Principles of Religion;* and as that Great Mathematician Mr. *Cotes* did in his excellent *Preface* to the later Editions of Sir. I. N.'s *Principia:* I mean for the advantage of Natural Religion, and the Interposition of the Divine Power and Providence in the Constitution of the World; His answer was, that He saw those Consequences; but thought it better to let his Readers draw them first of themselves: Which Consequences however, He did in great measure draw himself long afterwards in the later Editions of his *Principia,* in that admirable *Genera* [*sic*] *Scholium* at its conclusion; and elsewhere, in his *Opticks*. . . . Nor can I dispence with myself to omit the Declaration of his Opinion to me, Of the Wicked Behaviour of most modern Courtiers, and the Cause of it, which he took to be their having *laughed themselves out of Religion;* or, to use my own usual Phrase to express both our Notions, because *they have not the fear of God before their Eyes,* Which Characters being, I doubt, full as applicable to our present Courtiers, as they were to those of whom he apply's them long ago, is a Cause of great Lamentation.

Whiston concludes with the acid observation that in his experience, when honest Christians join a court, "I hardly ever find them to amend those Courts, but to be almost always greatly and fatally corrupted by them."[33] The "trimming" of Sir Peter King is a case in point.

This amazing tale suggests that Newton's disgust with the jesting freethinkers in the Court may have been one of the factors that finally led him to lend his own name and reputation to an argument designed to show that providence truly governs both the "world natural" and the "world politic." It also points out the independent nature of Whiston's (and Newton's) Whiggish political inclinations.

Whiston was always quick to criticize even his closest Whig friends when he detected a moral lapse. The eminent Whig journalist, Sir Richard Steele, upon the request of the secretary of the Society for Promoting Christian Knowledge, Henry Newman, greatly helped Whiston to get started in the business of public lectures in London after his expulsion from Cambridge. But Whiston retained his independence. Whiston records an encounter at Button's Coffeehouse with Steele in 1720 at which Whiston chastised Steele for giving a speech in the House of Commons in favor of the South Sea directors, even though Steele had written many articles against the South Sea scheme in his paper. Whiston states that Steele made the speech because without doing so he could not regain his former highly lucrative post as Censor of the Playhouse. Whiston states,

> I accosted him thus. They say Sir *Richard,* you have been making a Speech in the House of Commons, for the South-Sea Directors. He replyed, they do say so. To which I answered, How does this agree with your former Writing against that Scheme? His rejoinder was this: Mr. *Whiston* you can walk on Foot [do without a carriage], and I cannot. Than which a truer or an acuter Answer could not have been made by any body.[34]

Whiston's independence even from those of his Whig benefactors who had helped him most is only further underscored by his first political pamphlet, which is a kind of "modest proposal" for the elimination of the curse of all political factions. In his *Supposal: Or a New Scheme of Government. Humbly Offer'd to Publick Consideration by a Lover of Truth and Peace* (1712), Whiston shows that he is not a radical Whig but a moderate, interested primarily in a stable, orderly society unriven by the rage of party factions. Whiston's proposal is for the queen to appoint a privy council composed of men from both parties renowned more for their learning, integrity, piety, and wisdom than for being avid "Party-Men." The queen will then charge these privy councilors "to give Her always faithful Advice, according to the best of their Judgement, without all Regard to Party; and recommend none to Her for any Place, but for their Personal Vertue and Qualifications." The effect of their advice, rendered for the good of the queen and her people, will be that

> . . . the Church and State will be no longer confounded together; Courtiers will not influence Ecclesiastical Affairs, nor dispose of Church Preferments for Secular Ends; the Clergy will no longer intermeddle with Worldly Matters either in Court or Parliament; will not Preach up unlimited Obedience, to flatter the Prince; or Excuses for Disobedience, to humour the People; there will be no Persecution for Conscience sake; the Church will not have any of its Sacred Ordinances expos'd to contempt, by being Tests and Qualifications for Secular Employments: nor will any Person be tempted to Hypocrisy by being oblig'd to Occasional Conformity on those Accounts. . . Nor will such Things be enjoin'd which are forbidden, nor those made penal which are commanded in the Gospel.

Whiston reckons that such bipartisan administration, untainted by the rage of party strife, "*would be the Stability of our Times*" and bring about in England a true "Sion."[35] Whiston's disgust with the moral degradation in society produced by party factional strife ironically echoes that of his most devastating critic, the High Church Tory Jonathan Swift. In Book I of *Gulliver's Travels,* Swift's depiction of the source of rivalry between the two factions among the Lilliputians as a preference for low heels or round heels is directed mercilessly at Whig and Tory parties alike.[36]

Despite Whiston's spurning of radical (deistic) Whiggery, the Whig Court (as corrupt as any other court), and the Whig Party qua party, Whiston nevertheless shared the two most basic Whig doctrines urged by the Whig ideologues at the Sacheverell trial: toleration for religious dissent and the concept of the limited, contractual, but providentially granted nature of monarchy. Most of Whiston's writing about Arianism after his expulsion from Cambridge and during his trials in London is directly concerned with the issue of religious toleration.[37] In 1717, he unlimbered his Newtonian biblical interpretation in defense of the most basic of all Whig principles and the one that ensured the legitimacy of the monarchy after 1688. Whiston states that he wrote *Scripture Politicks* "fully for the Title of Princes, as not to be derived from Hereditary Right, but from the choice and Recognition of the People."[38]

Utilizing his now familiar interpretative principle that "the Obvious or Literal Sense of Scripture is the True and Real one, where no evident Reason can be given to the contrary," Whiston ransacks the Bible for texts to support his version of the Whig doctrine of limited, elected monarchy and the right of revolt if the monarch abrogates his or her contract. Whiston utilizes his Newtonian method of scriptural interpretation to elicit a latitudinarian theory of the nature of the contract between monarch and people midway between the radically republican, deist view – according to which the people have the power to elect and change monarchs by virtue of natural right – and the Tory view that the people always owe passive obedience to their monarch because he or she rules by virtue of divine right passed on by hereditary succession. Whiston is but another in a long line of Low Church divines (whose chief spokesman was Whiston's mentor within the Church, Bishop William Lloyd)[39] to urge a theory of divine right bereft of the Tory concepts of passive obedience and indefeasible hereditary succession. For them, the theory of divine right is modified to mean that the partnership between king and people is still divinely instituted but that God's providential direction of the succession operates through the mechanism of the choice of the people, not through heredity and right of primogeniture. Whiston did not believe that when James II fled the throne God went with him: the judgment of the people was providentially guided in choosing William as king.

Whiston employs his literalistic, Newtonian-inspired method, for example, to gloss Deut. 17, in which Moses prophesies the introduction of regal government among the Israelites and the nature of the rules for the qualification, election, and duties of those monarchs. The Israelites are not required by the law of God always to elect the eldest son of the royal family as their king. In their case, as in the English one, "the Succession of the Crown in any Royal Family was a Favour, which religious and obedient Kings, and those only, might hope for; and which Providence would generally bring about on such conditions; as instances, not of their strict Right of Succession, but of God's Blessing."[40]

This lesson is confirmed during the time of the judges (from the death of Joshua and the beginning of the regal government by Saul), when no sons claimed the right of hereditary succession. Instead, if they wished to succeed their fathers, they persuaded the people to elect them, as, for example, Gideon's son Abimelech did (Judg. 8–9). David's providential selection by God was accompanied first by the providential mechanism of his election by the people of his own tribe of Judah and then by a covenant of voluntary consent to his kingship with the rest of the tribes (2 Sam. 5:1–3). Jeroboam's election by the people after the death of Solomon and their rejection of Rehoboam (Ps. 7:20, 40:29–31) showed that the free election of the monarch "was so order'd by Divine Providence on purpose."[41]

As we saw in Chapter 2, Whiston gives example after example to show, literally, that just as God's providential mechanism of order in the natural world is gravity, in the coursing of the political world his providential direction is exercised through the election of the people. There is in fact, Whiston claims, only one instance he can recall in the whole Bible where succession proceeded among the Jewish kings "merely on Account of Primogeniture . . . and 'tis not a Divine, but a Human Preference, like that of the Royal Family of *Moab*." This was the settling of the succession by Jehosophat on his first-born son Jehoram, and it turned out badly for the Jews (1 Kings 22:42).[42]

The Bible shows that succession to the monarchy is by means of providentially directed election by the people. The Bible also contains instances of the right of the people to resist with force an illegal – that is, nonprovidential – succession. When King Amon was slaughtered by his servants, "The People of the Land Slew all them that had conspired against King Amon" (2 Kings 21:24) before electing, freely and providentially, King Josiah. Whiston notes,

> Here we seem to have a very particular Case; I mean an Instance of Supreme Authority, or of the Inflicting Capital Punishment in the taking away the Lives of Criminals by the People, before the Succession of a new King: And we have another plain Example, that the Kings of *Judah* did

Figure 1. In this eighth plate from Hogarth's *The Rake's Progress,* entitled *Scene in a Madhouse,* the Rake is shown being manacled for attempted suicide in the infamous Bedlam Hospital. On the wall behind him, another inmate of the asylum is sketching details of Whiston's method for determining longitude, including the firing of a star shell from a mortar. Swift and Arbuthnot no doubt appreciated this touch. From Joseph Burke and Colin Caldwell, *Hogarth: The Complete Engravings* (London: Thames & Hudson, 1968). Reproduced by permission of the Trustees of the British Museum.

Figure 2. Whiston appears with other "remarkable figures" in 1748 in this print from a drawing by Samuel Richardson published in *The Correspondence of Samuel Richardson* . . . , 6 vols. (London, 1804), vol. 3. Whiston, in the lower right-hand corner in his clerical collar, is striding vigorously and characteristically away from the crowd.

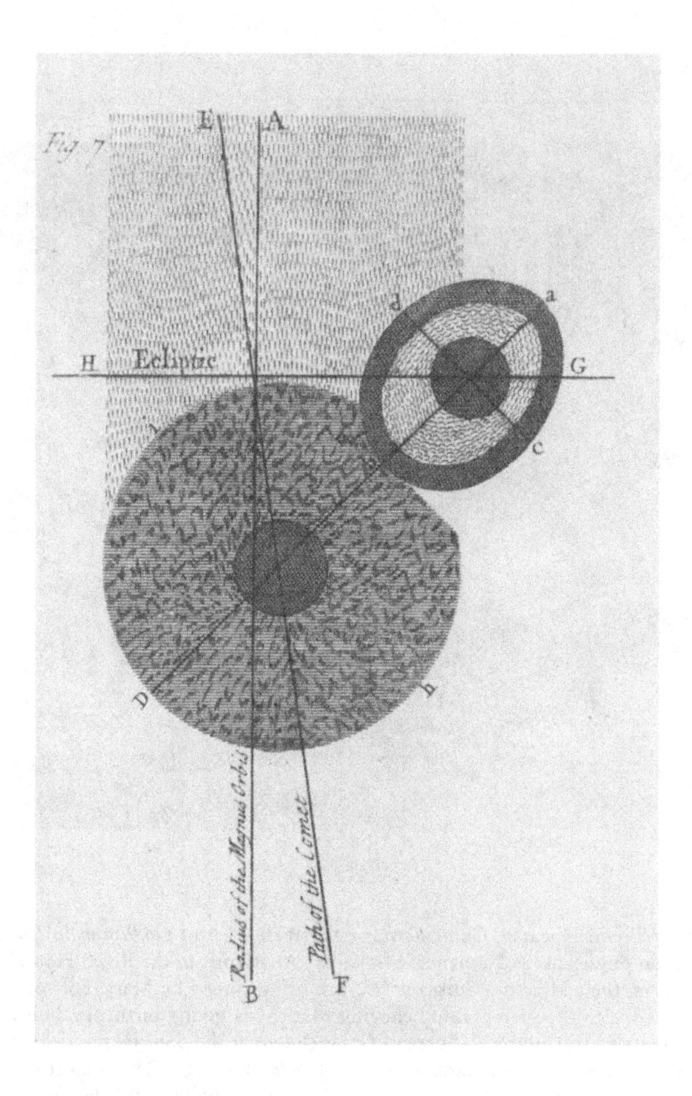

Figure 3. Whiston's deus ex machina. A comet passing along the line of trajectory *EF* squeezes the earth (*abcd*) into an oblate spheroid by the force of its gravitational attraction. This deformation fissures and cracks the earth's crust. While the earth is in the comet's tail, the watery vapors of that rarified atmosphere are precipitated as rain onto the earth. The weight of this rain plus the tidal effect in the dense fluid of the Abyss causes this fluid to make a violent ascent through the fissured crust to join the celestial rain from the comet's tail. Whiston's "clear, easie, and mechanical account" of the Noachian Deluge thus explains the "opening of the windows of heaven" and how "the fountains of the great deep burst forth." From William Whiston, *A New Theory of the Earth From its Original, to the Consummation of All Things . . .* (London, 1696). Reproduced by permission of the Syndics of Cambridge University Library.

Figure 4. In this print by William Hogarth, *Cunicularii, or the Wise Men of Godliman in Consultation,* such eminent London physicians as Nathanael St. André, Anatomist to the Royal Household (*A*) and Dr. Cyriacus Ahlers, their Majesties Surgeon (*C*) are being duped by Mary Toft of Godalming (the "Godliman" of the title), Surrey, into believing that she is giving birth to rabbits. Dr. John Howard, a male midwife from Guildford, Surrey (*D*), is shown at the door taking delivery of a baby rabbit from a local gamekeeper and complaining that it is "too big." The imposture was suspected by Dr. Sir Richard Manningham (*B*), the best-known male midwife of the day, when she "delivered" her seventeenth rabbit. After failing to repeat her miraculous feat in London, she confessed to fraud and served a short term in prison. Whiston believed her original claim and thought that she had been forced to recant her story, which he interpreted as the fulfillment of a prophecy in the apocryphal book of Esdras predicting such monstrous births. See Chapter 1, note 55, and Dennis Todd, "Three Characters in Hogarth's *Cunicularii* – and Some Implications," *Eighteenth-Century Studies* 16, no. 1 (Fall, 1982):26–46. From Joseph Burke and Colin Caldwell, *Hogarth: The Complete Engravings* (London: Thames & Hudson, 1968). Reproduced by permission of the Trustees of the British Museum.

not succeed of Course, by virtue of any Hereditary Right; but by the Choice and Recognition of the People.[43]

One of Whiston's most "fundamental" propositions is that the covenants and oaths providentially contracted between the people and their elected monarch cannot be abrogated, "even where great Inconveniences arise by the Observation of them." The people are duty bound to maintain their oath "and bear the Inconveniences of that Persons Government and Oppressions till his death; or till God, by some other Methods of his Providence, deprive him of that Power of Government; but no longer."[44] The usual barrage of texts is adduced to prove the providential nature even of revolution when that revolution is effected for the preservation of the worship of the true God.[45] Passive obedience and the denial of the right to resist a nonprovidential, heretical monarch are merely human "Ordinances and Decrees" and "one of the grossest Parts of *Popish* Tyranny."[46]

Gerald Straka is correct that this providentialist interpretation of the right to elect or evict a monarch falls, later in the century, before the law of nature in politics enunciated by Locke and developed by eighteenth-century deistic philosophies.[47] But the providentialist view of a king's election and the right to revolution constitutes the specifically Newtonian contribution to the Whig–Tory conflict in the first decades of the eighteenth century. The Newtonian universe does indeed reveal a providential father in heaven. His providential hand is revealed in the political history of the Jewish people through Newtonian biblical interpretation. God rules on earth in the same providential way he rules in heaven, and this fact legitimizes the Glorious Revolution. It was, simply, divinely sanctioned and revealed another aspect of the working of divine providence. Whiston does not rely primarily on the scientific theism of the Newtonian design argument to support his moderately Whiggish interpretation of the Glorious Revolution. He stresses, instead, the Newtonian insistence on the properly interpreted word of Scripture, directing his argument against both the materialistic, naturalistic, deistic republicanism of the more radical Whigs and against the High Church Tories, who, predictably, attacked it.[48]

II. Whiston's Newtonian biblical interpretation and radical Arianism

Just as they valued Whiston's help in the battle with the deist Collins about the validity of the project of interpreting Scripture, so the moderate Low churchmen probably also appreciated Whiston's Whiggish political polemics. But they drew the line when Whiston extended his method of literal scriptural interpretation into the demonstration of the nature of genuine – that is, "primitive" – Christianity. Nor did Whiston appreciate them. Whiston charges that the Low Church bishops are as detrimental to the proper ordering of church and state as the High Church Tories, because they hinder "our return to the Primitive Model, by wholly confining their Sacred and Ecclesiastical Enquiries to those, and only those Canonical Books of the New Testament which the *Popish* Churches and Councils have received and recommended to them: How Authentick soever any others were esteemed before the Papal Corruptions began."[49]

In the summer of 1708, Whiston became convinced, on the basis of his historical research and his reading of the *Apostolical Constitutions,* that Athanasius was a fraud and that the doctrine of the Trinity was a cruel hoax. As we noted in Chapter 1, he wrote to the latitudinarian archbishops, Tenison of Canterbury and Sharp of York, for advice on how to go about bringing this news to the attention of churchmen and laymen. He specifically inquired whether a printed book might be the best course of action. Whiston was fully aware of the political implications of publishing such an antitrinitarian, pro-Arian work and therefore of the possible argument that the proposed work ought to be suppressed on political grounds. Whiston knew that his publication of his views would provide an explosive political issue for the down-and-out High Church Tories to rally around, that of "Church in Danger." Whiston understood that the Tory high-flyers could argue that Godolphin's Whig ministry and the Whiggish Low Church bishops, by allowing publication of such a book, threatened the very fabric of established Church doctrine. In his letter to the archbishops, Whiston seeks to meet this argument for suppression of the publication of his discovery of the true nature of Christianity in advance:

> I am well aware that several political or prudential Considerations may be alledg'd against wither the doing this *at all,* or at least the doing it *now.* But then, if the sacred Truths of God must be always suppress'd, and dangerous Corruptions never inquired into, till the *Politicians* of this world should say it were a *proper Time* to examin and correct them, I doubt it would be long enough e're such Examination and Correction could be expected in any Case. I think my self plainly *oblig'd* in point of *Duty* to communicate my Collections to the publick Consideration; and therefore from this *Resolution* in general no worldly Motives whatever, by the blessing of God, shall dissuade me.[50]

Nevertheless, as noted in Chapter 1, the bishops tried to prevent him from publishing, even though Sharp himself agreed with Whiston's antitrinitarian position.[51] The bishops felt that the political situation was too highly charged for the implementation of Whiston's plan. The Sacheverell trial and the riots it caused shortly thereafter proved them correct.

Whiston plunged ahead with his application of his Newtonian method of scriptural interpretation to the demonstration that the trinitarian doctrine enunciated in the Athanasian creed was little more than a public "cursing" of true – that is, Arian – Christianity. That Whiston's biblical interpretation may be legitimately termed Newtonian in the context of his heartfelt Arianism is corroborated by the fact that the same method of interpretation that stands behind his preference for mechanical explanations of the causes of natural phenomena endorsed by Newton also stands behind his Arianism. Whiston's "Advice for the Study of Divinity" was begun in February 1708 and published in his *Sermons and Essays* of 1709. Ultimately it was used as evidence in the proceeding to banish Whiston from Cambridge for breaking the university statute against holding or teaching doctrines contrary to established Church doctrine. In this essay, Whiston uses his literalistic method to argue that the concept of the Trinity was not a part of the articles of faith of the first Christians and that it was added later by parties anxious to impose their system on "free enquirers" – that is, Newtonian biblical critics who can read the Scripture and understand God's meaning for themselves. He further examines the historical records to determine why it is that so many schisms have arisen even within the reformed churches. He concludes that it is because Christians have taken the wrong historical documents as their guide, preferring either the authority of commentaries or of councils and Fathers from the fourth and fifth centuries to the word of God from more primitive times, "literally" interpreted. Whiston recommends that one read the Bible and the original records from the time of the Bible rather than "Modern Books of Controversy, written by such as are to vindicate the Doctrines and Practices of that particular Sect or Church."[52] As for councils, Whiston notes that on any particular doctrine decided by the councils of the fourth and fifth centuries there are as many opinions as there are councillors. Consider the case of Arianism. An impartial observer

> ... will soon see Councils for the *Arians,* and Councils against the Arians, in abundance. He will presently observe the Imperial Power managing and over-awing the Councils. . . . He will but too plainly perceive the *common,* or rather *uncommon* Infirmities, Heats, and Passions of Mankind. He will find particularly in the *Arian* Controversy, not as in former Cases, One or Two single Men setting up their peculiar Opinions against the general Sense and Tradition of the Catholick Church; but the greatest and the wisest Men of the whole Catholick Church divided in their sentiments, and unable to come to any Certain Agreement.[53]

The schisms of modern-day Christianity, he says, had their roots in the same problem that stimulated the Burnet controversy. The problem is one of exegesis or interpretation. What is needed is a method, or guiding principle, that will not be too simpleminded and literal but that also avoids the opposite error of Burnet and the deists, who too willingly see in the mere fact of schism the invalidity of Scripture.

Whiston's Newtonian method of interpreting Scripture validates Scripture. His third exegetical "postulate" urges commentators to get to the true Scripture by ignoring later commentators and documents and by going back to the original documents from the time of Christ. He asks, "Would any man of common Sense depend on the Writings, Articles and Canons under Queen *Elizabeth,* as good Evidence of the Faith and Practice of the *English* Church in the Day of King Henry III or *Edward* I. . . . while he could have immediate Recourse to the Original Records of those Times[?]"[54]

Whiston and Newton, however, shared more than method. They also shared the Arian doctrine that the method elicited from the historical documents. According to the Newtonian view, Jesus was neither coeternal or consubstantial with the Father. Nor did Jesus bring together in his person a union of divine and human nature. Jesus was the created logos incorporated in human flesh, so that he, and not humanity, might suffer. For his perfect obedience to the divine plan he deserves worship, but not of the "idolatrous" variety that defines him – through the doctrine of the Trinity – as being of the same substance with God the Father. The corollary to this view is the theory of how the church perverted this original doctrine and, led by Athanasius, Jerome, Augustine and others, conspired to impose an idolatrous fraud upon Christendom. Whiston's Newtonian method of interpretation is used to establish that the Arian doctrine was the original church doctrine and that the trinitarian creed of Athanasius was a corruption purposely introduced into the records of church councils. Newton's views on this score, as we shall see, were discreetly expressed in his manuscripts and private papers.[55] Whiston, however, truly shrieked this news in the marketplace in his massive *Primitive Christianity Reviv'd* (1711) and in such tracts as *Athanasius Convicted of Forgery* (1712). This latter work was directed against Styan Thirlby, and it again combined Whiston's Newtonian exegesis and Arian doctrine. Thirlby had written a tract against Whiston's *Historical Preface to Primitive Christianity Reviv'd* (1711), demanding "full proof and undoubted Evidence" that Athanasius was in fact, as Whiston stated he suspected, the wicked instrument of a massive fraud. Whiston acknowledges that it is not possible to attain such certainty in a historical dispute of this type, even though, according to Whiston,

> I may have sufficient Foundation, vehemently to *suspect* a Person's Integrity, and to guard my self against him guilty of Knavery, in a Court of Justice: Especially when the World Knows, that the *Athanasians*

> have long ago dropp'd or suppress'd those opposite Arian Accounts, and many of those Original Records, which were likely to tend so much to their Disadvantage; and which yet must ingeneral be absolutely necessary in such a Method of demonstrative Evidence, or legal Conviction.[56]

Whiston argues that the ancient evidence nearest to the time of the Council of Nicea that is not derived from Athanasius himself shows that that council did not anathematize those who said Jesus was created, and hence of a like substance, but not the same substance, as God. The copies of the Nicene Creed and the anathemas they contained – possessed, for example, by Cyril of Jerusalem, Bishop Basil of Caesarea, Bishop Epiphanius of Salamis, and many others – omit to anathematize the Arians for maintaining that Jesus was created. The council did not censure this doctrine and, by omitting mention of it all, avoided its condemnation. Thus when Athanasius asserts, as he does over and over, that the Council of Nicea condemned or anathematized Arians who believed Jesus was created, "good Arguments" and "numerous Testimonies" reveal that he was lying.[57]

In a Newton manuscript entitled "Paradoxical questions concerning ye morals and actions of Athanasius and his followers," Newton also attempted, privately, to utilize ancient, uncorrupted texts to demonstrate that on several issues Athanasius was guilty of forging anathemas of Arian doctrines. In one *Quaestio,* Newton queries "whether Athanasius [& his friends] did not corrupt the records of the Councils of Nice & Sardica" (brackets in the original). Newton explains that the crucial term *homousious,* was understood in two senses: "that a thing is of the same substance with another or that it is of a like substance."[58] Newton concludes that the council never censored the Arian interpretation of the word – that Jesus is *like,* but not the *same* as, God – and that Athanasius falsified the council records to make it appear that they did.[59] Whiston states, in his openly published pamphlet, that he "received the first Intimation, and several of the Proofs" for this thesis from two "Learned and Judicious Persons,"[60] who were undoubtedly Newton and Clarke.[61]

Whiston's ouster from Cambridge for his Arian views, and his trials in London before the ecclesiastical Convocation and the Court of Delegates, do not constitute a mirror image of the Sacheverell prosecution. Sacheverell elicited support throughout the country for his High Church Toryism. Whiston, on the other hand, stood alone. The Low Church bishops, many of whom (according to Whiston) agreed privately with his Arian views and who had tried to warn him off, united with the high-flying Tory prosecutor Francis Atterbury (Sacheverell's defender) to prosecute Whiston's heresy. Eamon Duffy has pointed out that Whiston may actually have been set up as a target for the high-flyers at the Convocation to prevent them from effecting their primary goal of reviving the Church's influence in state affairs.[62] In this, the Low Church bishops may have been aided by the chief Tory minister Lord Harley, who, with Queen Anne, was determined to limit sharply the scope of the Convocation, a goal attained with the help of Anne's confessor, Archbishop Sharp of York.[63] The orthodox hostility to moderate Whiggery and Low Church policies that had been mobilized by the Sacheverell affair and expressed in the subsequent Tory landslide in 1710, and that was personified by the prolocutor of the Convocation, Francis Atterbury, was deflected onto William Whiston, a heretic who publicly espoused the views privately held by Newton.

At the same time that Whiston's trials for heresy were being conducted in London, he was undergoing extremely harsh personal attacks by the Scriblerians, the club of satirists composed of such arch-Tories and High Church literati as Jonathan Swift, John Arbuthnot, James Parnell, John Gay, and Alexander Pope, for his project of discovering the longitude (see Chapter 1). The viciousness of their ridicule of the longitude project seems to have been connected with Whiston's antitrinitarianism. Arianism was a crucial focal point of orthodox hostility, and to the High Church Tories such as Swift, Whiston's example implied an inevitable connection between the new philosophy of Isaac Newton, republicanism in its various forms, and Arianism. Schism in theology, party squabbles in politics, and sectarianism in science and philosophy all issued from the pridefulness of the Newtonians, who believed that their rationalism could fathom everything in heaven and on earth. Vain Newtonian rationalists engaged in inquiry ranging across the whole map of the world, human and divine, and even into mysteries heretofore concealed from the light of reason and historical inquiry. The Trinity is such a mystery, according to Swift, and reason cannot comprehend it.[64] Parnell wrote revolting doggerel against Whiston's project, just because he wanted to ridicule the "Confident Arian" (see Chapter 1, note 71).[65] Pope comments on this connection between natural Newtonian philosophy and the rationally based assault on established Church doctrine in the following ironic text from his "God's Revenge against Punning":

> But when Whoring and Popery were driven hence by the Happy *Revolution;* still the Nation so greatly offended, that *Socianism, Arianism,* and *Whistonism* triumph's in our Streets, and were in a manner become Universal.
>
> And yet still, after all these Visitations, it has pleased Heaven to visit us with a Contagion more Epidemical, and of consequence more Fatal: This was foretold to Us, First, By that unparallel'd Eclipse in 1714: Sec-

> ondly, By the dreadful Coruscations in the Air this present Year: And Thirdly, By the Nine Comets seen at once over *Soho-Square*.[66]

The contagion foretold by those decidedly Whistonian mechanisms of eclipses and comets is, of course, an outbreak of punning throughout society, which Pope then details as if it were divine retribution for a sinful nation that had lapsed into such heresies as Socianism, Arianism, and "Whistonism."

The *Examiner,* the Tory newspaper founded by Swift, reacted at the time of Whiston's trials to his vocal opposition to High Church doctrine, which was interpreted as an inevitable consequence of Whiggery. A Tory writer in the *Examiner* linked Whiston's Arianism with his Whig supporters, charging that Whiston led "a party, as *Arrius* did of old," and that because Whiston's party was hostile to orthodox Christianity of the Tory, High Church variety his cause was espoused by the Whigs.[67]

Whiston's antitrinitarian theology was thus associated with Whig republicanism. It was also intimately connected with Newton himself, who inspired its methodology and doctrine. But Newton was content to remain in the background, leaving it first to Whiston and then to Clarke to defend primitive, uncorrupted Christianity. Whiston's family background and his temperament had prepared him for his chosen task as the public defender for this aspect of the Newtonian program. Once Newton had shown him the way, Whiston was out in front of even Clarke in defending Arianism, in spite of the consequences to his career. Whiston saw himself, in fact, as a kind of new Luther whose appointed task was to revivify primitive Christianity and thus, to usher in the millennium. In his memoirs he describes a conversation he had with Dr. John Woodward about Whiston's fasting on Wednesdays and Fridays in accordance with the rules to that effect in the *Apostolical Constitutions*. Woodward chided Whiston for following vaguely Popish practices. Whiston replied that

> . . . had it not been for the rise now and then of a *Luther,* and a *Whiston,* he [Woodward] would himself have gone down on his knees to St. *Winifred* and St. *Bridget:* which he knew not how to contradict. 'Tis much safer to Keep the original rules of the gospel, than to invent evasions and distinctions how we may most plausibly break them, which is the way of the moderns perpetually.[68]

Westfall has raised the most important question about Whiston's various recollections of his relations with Newton, Clarke, and other convinced Arians friendly with Newton such as Hopton Haynes, Whiston's "intimate Friend" and Newton's assistant at the Mint. Westfall queries, "In

[Whiston's] recollections, one catches a glimpse – is it a true image or is it a mirage? – of one of the most advanced circles of free thought in England grouped around Newton and taking its inspiration from him."[69] It seems clear from an analysis of the similarity of method and doctrine in Newton's manuscripts and Whiston's books that Whiston's recollections are more of a true image than a mirage. The influence of Newton on Whiston's career, especially, cannot have escaped the notice of the vigilant defenders of orthodoxy at the time, for whom, as Olson has demonstrated, pride in scientific virtuosity led straight to both nonconformity in religion and revolution in politics.[70] My one quibble with Olson's very sound interpretation is his view that these High Church opponents – who attacked the overweening pride of the new scientists that led to such laughable excesses as the attempt to find the longitude by anchoring ships in a line across the wide Atlantic, and who attacked their repellent republican doctrines about the contract between prince and people and the right (natural or scriptural) of resistance, and who, finally, attacked their condemnation of the doctrine of the Christian church, including the Protestant churches, as the product of forgers and frauds – failed to attack "Newtonian science." Olson is correct in the narrow sense of "Newtonian science" but not in the broader meaning of Newtonianism. Newtonianism embraced all these facets and was understood by the orthodox Tories of the day (and by the latitudinarian bishops) to have these ramifications. When Arbuthnot, Swift, Pope, and Gay attacked Whiston's longitude scheme, his Whiggish politics, and his Arianism, they attacked what they saw as the inevitable consequences of the pridefulness of Newtonianism. They were sufficiently successful in their attacks on Whiston and Clarke so that Newton himself, having tested the wind through his self-propelled surrogate Whiston and through the more moderate Clarke, remained silent about his own Arianism, perhaps in the belief that until the millennium, "The wicked shall do wickedly: and none of the wicked shall understand; but the wise shall understand" (Dan. 12:10). Part of Newton's fear of Whiston in his latter years may have been caused by the knowledge that Whiston knew that Newton shared his heterodox views, and that if Whiston wanted to, he could stigmatize Newton with the same calumny that Whiston had had to put up with for his spirited public defense of primitive Christianity. As late as 1721, one of the foremost defenders of High Church orthodoxy, Daniel Finch, the second earl of Nottingham, after an exchange of pamphlets with Whiston on the topic of Whiston's antitrinitarianism, attempted to pass a bill in the upper House approving penalties against anyone who denied the divinity of Christ and the orthodox doctrine of the Trinity. The Whig ministry mustered more than enough strength to stifle this attempt to reverse the liberal social policies of the Whig government (the motion was defeated 60 to 31),[71] but Newton, himself a political appointee at the Mint, had seen many changes in government and attitudes. Whiston could have done great injury to Newton's reputation, but he did not. Nevertheless, he knew what Newton believed, and Newton, as a result, could not rest easy.

It seems fair to say that Newtonianism – with its calm, rational assurance that a properly historicist interpretation of the original Christian documents revealed Athanasius and his trinitarian creed to be anti-Christian frauds – was in fact tarred with the brush of negative controversy when Whiston and Clarke, two of Newton's closest disciples, were pilloried before the Convocation of 1710–14 as heretics, despite Newton's own reticence and timid reluctance to enter the fray. Implicit in this is a criticism of Jacob's thesis that Newtonianism triumphed in 1720. This thesis depends on Jacob's unstated principle that Newtonianism and latitudinarianism were identical. But there is a distinction. Low Church, Whig latitudinarians united with High Church Tories to brand this aspect of Newtonianism as heretical. Whiston's prospects for advancement never recovered from the stigma of his trials in the period from 1710 to 1714. Clarke never advanced to any high Church position, even though in Whiston's view, Clarke had trimmed his Arian sails just enough to gain some preferment.[72] The Low Church bishops, though perhaps in private agreement with the Arian doctrine, united with the orthodox Tories on this point of doctrine and succeeded in quashing it so completely that Newton and Clarke were intimidated. Only Whiston remained publicly true to Newton's own privately nurtured beliefs, at great personal cost. Latitudinarianism and a degree of tolerance for Dissenters such as Whiston certainly triumphed by 1721, when the Whigs crushed the earl of Nottingham's bill of sanctions. But the Arian portion of the Newtonian system did not triumph. It can have done this aspect of the Newtonian cause little good when such deists as John Toland defended Whiston.[73]

III. Great expectations: the millennium and Whiston's Newtonianism

Whiston's millennial expectations, along with his other demonstrations of God's providential activity in human affairs, are ultimately founded upon his literal, historicist, Newtonian method of scriptural exegesis. Whiston never claims to be a prophet himself but only one who is able to connect the events of daily human and natural life to prophetic predictions and to show their literal fulfillment. Lecturing on the way in which current political events (such as the 1745 Jacobite rebellion) and natural phenomena (such as the London earthquakes of 1750) fulfill God's providentially revealed prophecies became Whiston's "peculiar business" in the last decade of his life.[74] Whiston did not arrive at this view in his dotage, however. It had formed an integral element in his thought since his *New Theory of the Earth* (1696), which described the end of the world in physical terms, and his *Essay on the Revelation* (1706), his most detailed work on the subject, which he continually revised. In 1712, he informed Prince Eugene of Savoy, who was visiting London, that the prince's victory over the Turks at Corfu and the subsequent Peace of Carlowitz had fulfilled Rev. 9:15. The prince sent Whiston fifteen guineas and a note thanking him for bringing it to the prince's attention that he "had the honour to be Known to St. John."[75]

Whiston's method for understanding the future events revealed in millennial prophecies bore a strong resemblance to Newton's. Both writers insist that the fulfillment of as yet unfulfilled prophecies cannot be predicted yet with sufficient accuracy, even by the scientific exegete. Only after the fact can the interpreter see how a particular event fits into the prophesied sequence with a high degree of probability. Whiston and Newton claim that interpreting the progressive unfolding of the millennial prophecies, such as the case of Prince Eugene's defeat of the Turks, clear and certain light is gained only "by degrees," with the passage of time.[76]

In his youth and middle age, Newton placed the day of final judgment at least two centuries in the future. Toward the end of his career, Newton moved the date at least three centuries into the future, if not farther off.[77] It is in Whiston's lifelong insistence upon the imminence of the millennium that he diverged most from his mentor. As early as 1706 they had quarreled seriously about how to reckon the chronology in the prophecy of the seventy weeks in Daniel, and Whiston states that his dogged insistence on his own views in such matters ultimately caused the rift between them.[78] Nevertheless, although they ultimately arrived at different calculations and expectations about the nearness of the return of the Messiah, they shared the same basic method of Scripture interpretation.[79]

Whiston, like all biblical interpreters, based his millennial expectations on the time periods mentioned in the prophecies in the books of Daniel and Revelation and the fates predicted there for the political kingdoms of the world. Two of the most crucial time references are "time, two times, and half a time" (Dan. 7:25) and the mention of 1,260 days in Revelation (12:6). Political states are often symbolized in these prophecies by beasts and horns of beasts, though this is not always the case, as, for example, in Daniel's vision of the metallic man (Dan. 2). In unraveling these prophetic predictions, both the identity of the nation and the length of time it will last must be determined.

In Chapter 7 of Daniel, the famous image of the four beasts rising from the sea – a lion, a bear, a leopard, and a monster – represent, Whiston and all other biblical interpreters of the day believed, the rise of the four great world empires, which is synchronous with Daniel's vision in Chapter 2 of the metallic man. The final, monstrous fourth beast, "which was diverse from all the others, exceeding dreadful" (Dan. 7:19) possessed ten horns on its head. These horns symbolize the ten nations that spring forth out of the fourth beast (or nation) and are synchronous with the parallel beasts of Rev. 13 and 17. Of these ten horns that issue from the fourth beast, one in particular, "a Little Horn," shall arise and assert dominion over three of the others (Dan. 7:24.) Whiston identifies the "Little Horn" of Daniel with Saint John's seven-headed beast (Rev. 17:9). Both, he says, refer to the Papacy, which has its seat in the seven-hilled city of Rome and which conquered three of the ten kingdoms that followed the Roman Empire, those of the Ostrogoths, the Lombards, and the Heruli. The Papacy, in good Protestant fashion, is also identified as the whore of Babylon of Rev. 17: "This *Beast with seven heads and ten horns*, is the *Roman* Empire after it was become *Christian*, in an Idolatrous and Persecuting Domination. Or in other words Babylon the *great Harlot that rides upon the Beast*, is not *Rome Pagan* but *Rome Christian,* fallen into an anti-Christian Tyranny and Idolatry."[80] The Papacy is the apostate, anti-Christian Roman Catholic Church, whose very name combines the secular and the sacred. The anti-Christian papal compromises with paganism and secular political power are a form of spiritual adultery akin to the moral "adultery" of Israel (Jer. 3) and Judah (Ezek. 6), hence the symbol of the harlot.

Having identified the players in the divine cosmic drama, Whiston is anxious to determine, so far as it is possible to do, the length of the play. How long until the chaste woman "clothed with the sun" (Rev. 12:1) returns from her banishment in the wilderness (Rev. 12:6) to displace the anti-Christian harlot? In short, when will Antichrist fall and the son of the chaste, sun-clad, apostolic church return? To answer this question, Whiston turns to the prophetic time periods.

Whiston, as we have seen, following Joseph Mede, Isaac Newton, Henry More, and all other biblical interpreters of the period, holds the basic hermeneutic principle that a day in biblical prophecy refers to a year in modern terms. Saint John's vision, in conjunction with this principle, provides one of the most important time prophecies: "And the woman fled into the wilderness, where she hath a place prepared of God, that they should feed her there a thousand two hundred and threescore days" (Rev. 12:6). This 1,260 days (or years), according to Whiston, is the "grand period" of time prophecies. At the end of this period, the persecuted, pure, apostolic church will be delivered. This period commenced when the ten anti-Christian kingdoms (represented by the ten horns that grew out of the monstrous fourth beast) succeeded the fourth beast, the Roman Empire. That occurred in A.D. 476, when the last Latin Roman emperor was replaced by the Ostrogoths. Adding 1,260 to 476 yields the year 1736. The fast-approaching year 1736 will therefore be the year when the process of the ten kingdoms casting off the yoke and uniting to defeat the beast commences. Exactly when the job will be completed and the pristine church restored cannot be determined precisely, because it is governed by an indeterminate time prophecy, the "time, two times, and half a time" (Dan. 7:25).

Only after the passing of the amount of time described in this phrase from Daniel will the "Little Horn" be completely vanquished, but how long, precisely, is this period? For Whiston, the term "time" in this passage refers to one year that is composed of 360 prophetic day-years. "Times" refers to two years, consisting of 720 prophetic day-years. "Time and times" thus equals 1,080 prophetic day-years. But the phrase "dividing of time" is vague. Whiston "supposes" that this period will be one-half a year, or 180 prophetic day-years, which, when added to 1,080, naturally equals 1,260 prophetic day-years.

This 1,260 must be added to the year A.D. 606, when Boniface III, the sixty-fifth bishop of Rome, gained the new title of "*Universal Bishop or Head of* the Church."[81] This was the point when the papal "Little Horn" gained mastery over the ten anti-Christian kingdoms of Rome, and therefore it is from this point that the "time, two times, and half a time," or, in Whiston's interpretation, 3½ times or 1,260 prophetic day-years, must be measured. Adding 606 to 1,260 gives 1,866.

Thus Whiston believed that between the years 1736 and 1866, a gradual wearing down of anti-Christian authority would occur as each one of the ten successor kingdoms in Europe defected from Antichrist.[82] Whiston adds, "within which Latitude [1736 to 1866], till farther light appears, it is now justly left undetermined."[83]

Though the exact time is indeterminate, Whiston believes that two "great Mutations are to be expected" soon after 1736, as predicted in prophecy. First, the "great Trumpet" shall sound, recalling the Jews to their native land (Isa. 27:13). Once restored, the "Sanctuary" or Temple must be cleansed and rebuilt (Dan. 7:13) and the Gentiles evicted from it (Rev. 11:2). Second, soon after 1736 Whiston expects "a great earthquake" in which a tenth part of a major city will fall and seven thousand men die, so that the survivors "were affrighted, and gave glory to the God of Heaven" (Rev. 11:13).[84] Whiston's lectures, with his model that was an exact reproduction of the Temple, fits into this context, and so does his concern for knowing the location of the Lost Tribes of Israel. He proposed various locations for the tribes in the course of his life (in 1724 he thought it likely that two of them were in Arabia) but was always an exponent of experimental verification:

> Nor would it be unworthy of Christian Princes to send thither Travellers so skilful, and well supported, and well provided with Necessaries, that what I here propose as only a *Probable Conjecture,* might be determin'd by *certain Accounts:* Which we know are greatly wanting in all these inward parts of *Arabia,* as in those of *Africa* and *America.*[85]

Whiston's interest in the exact location of the Lost Tribes seems to have been purely the result of intellectual curiosity. He formulates no plan for fetching them back to Jerusalem, nor does he feel that it is necessary for them to be converted.[86] The moving cause of the Jewish restoration would be God acting in human affairs. Whiston writes:

> 'Tis true, some things are imply'd in the Belief of the future Resettlement of the *Jewish* tribes, which cannot be explain'd without somewhat Extraordinary in the Case; without a sort of Divine and Supernatural Conduct. But this is so far from an Objection, that it implies no more than what was ever true of that peculiar People of God, in all the former Ages of their Commonwealth: About which God did all along exercise an extraordinary, nay a miraculous Providence continually, as every where obvious in the Sacred History. As likewise is it equally obvious in the Prophecies, that the same Extraordinary and Miraculous Conduct is to be again exercis'd towards them in this their Final Restoration and Resettlement in their own Land.[87]

Just as he is interested in the Restoration of the Jews, Whiston is alert for the coming earthquake. We shall see in Chapter 5 that Whiston's expectation of a devastating earthquake led him to view the London earthquakes of 1750 as providentially given signs of impending doom.

In his writings concerning the millennium, Whiston's basic interpretative method and the expectations based on it are similar to those of Isaac Newton. Both adopt Joseph Mede's prophetic day-year, wherein "days" mentioned in prophecy are counted as years; "weeks" as seven-year periods; and "months" as thirty years. Newton and Whiston also agree that apostasy reached its apex around A.D. 606 and that this date must be used for calculations about the ascendancy of the "Little Horn" in Daniel. Both agree, too, that 1,260 prophetic day-years is the critical prophetic time period during which apostasy will reign. Consequently, both locate the start of the millennium no later than around 1866, though for Whiston it might come much earlier and though Newton later revised the date past the year 2000.[88]

The most startling point of similarity in their views, however, is the confluence between the heartfelt Arianism of the two men and their millennial hopes. Neither Whiston nor Newton believed that once the millennium had arrived either the Church of England in its present corrupted form or any other Protestant sect that espoused the perverted trinitarian doctrine imposed by agents of the Whore of Babylon would be the restored apostolic church brought back from the wilderness. That true and pristine church would be composed of the primitive Christians who refuse to worship false idols. Whiston's reference to himself as a new Luther and his (from a twentieth-century perspective, perfectly quixotic) Arian proselytizing must be viewed in the context of a man who sees himself as an emissary of the pristine apostolic church, still in the wilderness, preaching quite near the end of time to the wise who would understand, while the wicked run to and fro doing wickedly. When he had to resign from the Society for the Preservation of Christian Knowledge because of his public Arianism, his letter of resignation made explicit how he could withstand the immense social and economic pressures brought on him to remain silent and not to rock the boat of established Church doctrine. He writes that he shares the society's goal of "doing good" but must nevertheless resign and that

> till those important things, I have to propose to the Christian world, be so thoroughly examined, that I may stand justified before all good men, and they may see it necessary to join my designs with those which they are already engaged in, in order to a thorough hastening the coming of our Saviour's Kingdom of peace and holiness. This, I verily believe, will be found necessary in no very long time. But since it is not in that state at present, and suspicions and jealousies may easily arise in the mean time, I do hereby take my leave of the society; begging of God . . . to open the eyes of the Christian world, to see, believe and practice exactly according to the revelation of his Son.[89]

By going public with his Arianism and his millennial expectations in his ferocious, uncompromising way, Whiston, to a greater degree even than Clarke, deserves the title of "the Honest Newtonian." He saw the world as a wicked and evil place. Most of the people he knew, even those few who were devoutly religious, ignored his voice from the wilderness and consequently, in his view, bore the mark of the beast. In the evil of the secular and corrupted state and church in which Whiston found himself, he could not be conjured into silence, the more so because he possessed the key to revealing the truth in his Newtonian method of scriptural interpretation. He was content to be banished from Cambridge to the wilderness of London to preach the one providental truth – in science, politics, and religion – there to await the providential overthrow of the existing evil order of Antichrist and the establishment of the reign of Christ and the saints who had remained true. As we shall see in Chapter 5, such dependence on the sudden, providential, premillennial reemergence of God into human history through the Second Coming of Jesus ran counter to the Enlightenment view of slow, steady human progress to a secular utopia. This aspect of Newtonianism had fitted more comfortably into the preceding century, and in trying to keep it active into the year 1750 Whiston achieved the reputation of a lunatic.

IV. Conclusion

This interpretation of the political and theological elements of Newtonianism illustrated by William Whiston forces a reconsideration of the Jacob interpretation. Jacob does not distinguish between the latitudinarianism of 1720 and the Newtonianism of 1720, the date by which, in her view, Newtonianism (or latitudinarianism) had "triumphed" over Tories and deists alike. Jacob's interpretation naturally arises as a result of abstracting out of Newtonianism precisely those elements most relevant to the structure of political and church society at the time. By emphasizing the order and stability of nature revealed by the Newtonian design argument and by ignoring the scriptural basis that the Newtonians used to legitimize revolution in politics (as with the Glorious Revolution) and in church doctrines (as with Arianism and millennialism), Jacob achieves a sanitized Newtonianism that is in fact identical to the outlook of some of the latitudinarian bishops of the Low Church. But it is not the Newtonianism of Whiston and Newton. Jacob does not mention Newton's (or Whiston's, or Clarke's) Arianism and seems to believe that by 1720 the Newtonians themselves no longer were convinced of the future millennium. Westfall's analysis of Newton's manuscripts and Whiston's published works shows, however, that Arianism and millennialism remained vital to Whiston and Newton. The design argument was certainly important, but the order and stability of the design argument applies to *this* heaven and *this* earth only, and even in *this* world Scripture provides reasonable grounds for providential revolution against an overweening prince. Both Whiston and Newton project as integral components of their thought the coming of a new heaven and a new earth, though disagreeing about the imminence of the event. Whiston even saw his preaching of his Newton-inspired Arianism as an event that would help in bringing about the downfall of the "Little Horn."

Jacob, in describing the "triumph of Newtonianism," sets great store by the Boyle Lectures as a mechanism for the dissemination of her orderly, stable, design argument Newtonianism. In discussing them she does cite Whiston's 1707 Boyle Lectures as illustrative of the Newtonian concern to demonstrate God's providence by showing how God had fulfilled prophetic predictions in the past. But Jacob also notes with puzzlement Whiston's *Essay on the Revelation* (1706), with its explicit millennialism, that precedes the socially acceptable catalog of already fulfilled historical prophecies in Whiston's 1707 Boyle Lectures. She accounts for this apparent change of tone by arguing that Whiston, too, trimmed his sails on the occasion of the Boyle Lectures and retreated from his millennial views, probably as a result of the "furore" caused by the French Prophets.[90] But, as I have shown in Chapter 3, Whiston despised the French Prophets and considered his Boyle Lectures as a development of an important aspect of Newtonianism, the belief in the literal fulfillment of the Scripture prophecies of the Messiah in the person of Jesus. But Whiston's belief in the fulfillment of scriptural prophecies and his belief that the millennium was near were not separate concerns. The high degree of certainty with which he awaited the fulfillment of the prophecies about the millennium had its epistemological basis in the Boyle Lecture demonstration of God's providential fulfillment of scriptural prophecies. We have reasonable grounds for believing in the fulfillment of prophecies whose times are not yet come on the basis of the fulfillment by God of those historical prophecies whose times have come and that are plainly shown in scriptural history. Whiston's *Essay on the Revelation* and his Boyle Lectures, the topic for which was suggested by Newton, are part and parcel of the Newtonian program and must, as Whiston insists, be taken together. It cannot be accidental that Newton suggested a catalog of already fulfilled prophecies as the topic of the 1707 lectures one year after the publication of Whiston's most expressly millennial book. Perhaps he hoped that Whiston's lectures would move the audience to a consideration of Whiston's millennialist tract. If not, the wise would understand.

It appears, then, that Newtonianism and latitudinarianism were not merely distinguishable but in active opposition. The one point of agreement between a high-flyer such as Atterbury and a Low Churchman such as Wake was that Whiston, a heretic, must be stopped.

5
DEISM AND DIVINE PROVIDENCE IN WHISTON AND NEWTON

IN the preceding chapters, I have shown why much of Whiston's anti-deistic writings may justly be termed Newtonian. In his demonstration of how particular historical events depicted in Genesis properly accord with scientific principles, in his "literal" reading of prophetic texts that he thought had accurately predicted the future course of determinate historical events, in his millennial expectations based on his "literal" interpretation of unfulfilled prophecies of the future, and in his fervent anti-trinitarianism, Whiston shared Newton's basic point of view. Because all of these views are based on the first two principles laid down in Whiston's introduction to *A New Theory of the Earth*, I have argued that Newton shared these basic principles of interpreting Scripture and have termed this approach Newtonian. On these points of theology, Frank E. Manuel is correct when he states that "because they are so forthright, William Whiston's works cast important light on the hidden intent and meaning of similar writings by his great contemporary. Where Newton was covert, Whiston shrieked in the marketplace."[1]

But Whiston is finally not a complete Newtonian, in spite of shared general principles of biblical interpretation and consequent shared theological positions. Whiston is a Whistonian, not a Newtonian, on the point of his third postulate of interpretation: "What Ancient Tradition asserts of the constitution of Nature, or of the Origin and Primitive States of the World, is to be allow'd for True, where 'tis fully agreeable to Scripture, Reason, and Philosophy." By the mid-1680s Newton had grown skeptical about making the Bible the criterion for the historical and chronological accounts in all ancient documents and non-Judaic historical traditions.

As Richard S. Westfall has clearly shown, in Newton's single most important theological manuscript – his "Theologiae Gentilis Origines Philosophicae" – we see Newton the historian at work on the historical and philosophical origins of Gentile theology. In his historical account, Newton treats ancient pagan sources as equally authoritative with the Bible. On several historical points, Moses' historical account of the early ages soon after the Flood are thought by Newton to be inadequate, and he corrects them in the light of such pagan sources as Berossus the Babylonian and Sanchuniathon the Phoenician. The general equation of the authority of Moses with these pagan ancients and the many analyses Newton makes of the "original true religion" in which such sources are given credence over the Mosaic account has led Westfall to ascribe to Newton a kind of gentle deism.[2]

Whiston, on this one point, diverges from the Newtonianism of Newton himself and upholds the traditionalist view that the Bible is the one infallible source of ancient history. This point of divergence between Whiston and his mentor underlies Whiston's successful criticisms of Newton's historical chronologies, as Manuel has so clearly shown.[3]

In what follows, I wish to elaborate on the nature of Whiston's dispute with Newton regarding ancient history and its sources while locating each man's attitude within the context of the deist movement. In Section I of this chapter, I therefore illustrate precisely to what degree Whiston and Newton are antideistic. The Newtonian method of scriptural interpretation, stated in Whiston's first two postulates, reveals, in Whiston's steady application of it, the providence of God in both the natural world of creation and in every aspect of human society and politics, past and future. Against Blount's mocking interpretation of Burnet and against Collins's equally mocking dismissal of prophecies, Whiston utilizes his distinctly Newtonian method of interpretation to reveal the providentially voluntarist God of Newton's "General Scholium" (1713).[4]

Divine providence purposefully orders everything on earth and in heaven. In order to prove this to his own satisfaction, a Newtonian biblical interpreter such as Whiston must properly interpret the Genesis creation story, show the fulfillment of messianic prophecies, describe the nature of the best form of government, determine the original doctrines of primitive Christianity, and reveal how current events fit the predicted pattern of the unfulfilled prophecies. In Section I, I analyze how Whiston and Newton grappled with the problem of synthesizing general and special providence. Newton and Whiston stood shoulder to shoulder in their efforts to defend the specially provident God of the Bible against the ridicule of the deists, a project made necessary, ironically, by the success of Newton's version of the design argument, which seemed to demonstrate only God's creative general providence.

In Section II, I trace Whiston's attack on Newton's deistic historical chronology. Behind Newton's chronology and his writing on ancient history is his attitude that the Bible is merely one source among many others and must occasionally be supplemented by superior pagan sources. Newton's attitude about the specific portion of Scripture that gives straightforward historical accounts of ancient events, especially those that occurred around the time of the Flood, is in fact deistic, as Westfall has pointed out, although Newton never shared the caustically mocking spirit regarding Mosaic history shared by more radical deists. Whiston violently opposes any interpretative method that invalidates even a portion of Scripture. Whiston's third postulate thus underlies his criticism of Newton's chronology.

In Section III, I show the philosophical problem that lies behind Whiston's and Newton's antideistic project of demonstrating God's providential control over his creation. This philosophical problem revolves about the distinction between God's two separate kinds of providential power. "General providence" refers to God's action in the original creation of nature. In the beginning, God created the material frame of nature, *and* he structured it to function in obedience to the laws of nature, which he also created at that time. In contrast to this original creative act of general providence – revealed so clearly in the design argument – is "special providence." Special providence refers to a particular act of providential intervention in natural or human events that cancels or contravenes the ordinary course of natural law. The historical record of the Bible reveals in accomplished miracles and fulfilled prophecies God's continued beneficent care for and governance of the "world natural" and the "world politic" through acts of divine special providence. The philosophical problem arises when the new scientists increasingly elaborated the natural laws that guide nature's operations and revealed a seemingly undeviating "clockwork" mechanism. Such a mechanism well illustrates God's initial creative act of general providence but is difficult to reconcile with the miracle-working, prophecy-fulfilling, specially provident God found in the Bible. David Hume's attack on any attempt to conjoin God's special and general providence is finally based on a rejection of the Bible's authority as a divinely inspired, and hence certain, source of ancient history. Hume's comprehensive religious skepticism thus grows out of a rejection of Mosaic authority similar to Newton's own, a development that would have horrified Newton but that follows naturally enough from his own position.

Section IV serves as a brief conclusion to the book.

I. Balancing God's general providence against his special providence in the apologetics of Whiston and the early Royal Society

The Christian virtuosi and the Royal Society of Isaac Newton

From the beginnings of the Royal Society, its founders argued that natural philosophy leads man to God, not away from him. The early Christian virtuosi who were prominent in founding the Royal Society generally insisted that their inquiries into the operations of the laws of nature revealed strong evidence of creative general providence and a specially provident deity capable of direct intervention into and disruption of the orderly operation of the generally provident machine of nature.[5] Bishop John Wilkins, Sir Walter Charleton, Robert Boyle, and Bishop Thomas Sprat all argued that nature reveals both general and special providence, although they also generally agreed with Wilkins that it simply is "not reasonable to think that the universal laws of nature, by which things are to be guided in their natural course, should frequently or upon every little occasion be violated or disordered."[6]

Newton, as president of the next generation of the Royal Society, and William Whiston (who was not a member) also labored to preserve both kinds of divine providence. As Whiston makes clear, this project was necessitated by the success of the Newtonian versions of the design argument in demonstrating God's creative general providence.

The first person to utilize Newton's law of gravity to demonstrate God's general providence was Richard Bentley, in the 1692 Boyle Lectures.[7] In the second edition of the *Principia* (1713), Newton himself posited the creator-architect deity of general providence in his famous "General Scholium." But in addition to the generally provident celestial clockmaker, the Newtonians attempted to preserve a specially provident God who daily superintends his creation down to the last sparrow's last flight. In the system of Newtonian theism worked out by Newton and his followers, God displays his special providence by miracles *and* prophecies.

Miracles

The original creation of gravity demonstrates God's general providence, but its continuous operation since that time reveals his special providence. God's sustained preservation of the order of nature and natural laws demonstrates divine special providence, because of the very nature of gravitational attraction. Newton even claims that "a continual miracle is needed to prevent the sun and fixed stars from rushing together through gravity."[8] This point is echoed by Clarke[9] and Whiston. Whiston is especially clear in his assertion that the very fact of nature's continuous operations in accord with natural law proves divine special providence: "'Tis now evident, that *Gravity*, the most mechanical affection of Bodies, and which seems most natural, depends entirely on the constant and efficacious, and, if you will, the supernatural and miraculous Influence of Almighty God."[10]

When Newton and Whiston declare that the daily operation of gravity is a miraculous effect of God, they mean that in obeying natural laws physical objects continually exhibit signs of God's special providence. But if all instances of obedience to the laws of nature are miraculous, then the traditional sense of miracles as a special denial or negation of the laws of nature is set aside. Obeying the laws of nature becomes, in this Newtonian interpretation, the supernatural effect of God's specially provident universal rule. The commonest natural event is itself a miracle. As Whiston says,

> I do not know whether the falling of a Stone to Earth ought not more truly to be esteem'd a *supernatural Effect,* or a *Miracle,* than what we with the greatest surprize should so stile, its remaining pendulous in the Open Air; since the former requires an *active Influence* in the first Cause, while the latter supposes *non-Annihilation* only.[11]

Newton and such Newtonian disciples as Whiston tend to discount the traditional conception of miracles held by Boyle, Wilkins, and the other founders of the Royal Society. A miracle is not a transgression of natural law. Rather, the sustained operation of natural law is itself a miracle and illustrative of divine special providence. Newton observes, and Whiston echoes, that miracles in the traditional sense are simply misunderstandings on the part of the vulgar. They "are not so called because they are the works of God, but because they happen seldom and for that reason excite wonder."[12]

It is their agreement about the impossibility of a cessation or interdiction of natural law that leads Newton and Whiston to their similar hermeneutic methods for interpreting such scriptural history as, for example, the creation story. Moses was popularizing; that is, he was "accomodating his words to the gross conceptions of the vulgar,"[13] not giving a "philosophical" account in terms of mechanical causes.

In providing scientific explanations of such apparent exceptions to the laws of nature as the creation of the earth and the Flood, Whiston went so far that John Keill – another Newtonian, and Savilian Professor of Astronomy at Oxford – attacked him for excluding special providence in his excessive reliance on "mechanism." Keill isolated himself from his more euhemeristic fellow Newtonians such as Whiston, Bentley, and Halley when he remarked that, like Boyle, he believed in miracles in the traditional sense because very often the only possible explanation for a natural phenomenon is a specially provident suspension of natural law by the deity:

> ... altho' Mr. Whiston has been pleas'd to ridicule my fondness for Miracles, yet since all the natural causes he has assign'd are so vastly disproportionate to the effects produc'd, he may at last perhaps be con-

vinc'd that the easiest, safest and indeed the only way is to ascribe 'em to Miracles.[14]

Finally, however, even Whiston equivocates about the miracles of Jesus Christ and, in the manner of his predecessors, accepts the possibility of a specially provident intercession abrogating natural law.[15] When a natural – that is, a "Nice and Philosophical Account" (to use Whiston's phrase) – is not possible, as in reliably reported stories such as that of John Duncalf or correctly interpreted scriptural accounts of the miracles of Jesus, Whiston accepts them as genuine transgressions of natural law.[16]

Samuel Clarke makes it clear in his correspondence with Leibniz that the chief issue between Leibniz and the Newtonians is that Leibniz tends to eliminate a specially provident God. The very perfection of Leibniz's (admittedly generally provident) mechanical system excludes "*Providence* and *God's Government* in reality out of the world."[17] Like Whiston, Clarke believes in the possibility of miracles as transgressions of natural law by a specially provident God. In his own Boyle Lectures of 1705, Clarke writes that a miracle

> ... is a work effected in a manner unusual or different from the common and regular method of Providence by the interposition either of God Himself, or some intelligent agent superior to man, in the proof or evidence of some particular doctrine or in attestation to the authority of some particular person. And if a miracle so worked be not opposed by some plainly superior power nor be brought to attest to a doctrine either contradictory in itself or vicious in its consequences – a doctrine of which kind no miracles in the world can be sufficient to prove – then the doctrine so attested must necessarily be looked upon as Divine, and the worker of the miracle entertained as having infallibly a commission from God.[18]

The Newtonians thus use miracles in two senses to demonstrate God's special providence. First, God's continuous act of preserving the natural world is specially provident; Whiston urges that the very operation of natural law is miraculous. But when pressed by Leibniz about the Newtonian reduction of God to an inferior clock repairman or when considering reliable scriptural reports of phenomena for which there is no mechanical explanation, the Newtonians accept miracles in the traditional sense as proof of God's continuous special providence.

Prophecies

Ultimately, however, fulfilled prophecies become an even more important tool than miracles for demonstrating God's special providence in the apologetics of the "first" Royal Society.[19] Fulfilled prophetic predictions tidily demonstrate the messiahship of Christ and reveal, once and for all, the hand of a specially provident deity continuously active in his creation. At the same time, fulfilled prophecies seem to avoid the dilemma of how it is possible for the generally provident order of natural law to be suspended in a specially provident miraculous act, and also the problem of spurious miracles.

Locke, Newton,[20] and Whiston use the many instances of apparently fulfilled prophetic predictions recorded in the Bible to supplement the design argument of natural religion, to round it out and draw from the total package the special, as well as the general, providence of God. As we have seen, Whiston, at the original suggestion of Newton, presented the 1708 Boyle Lectures on the topic of *The Accomplishment of Scripture Prophecy*. Newton, in his *Observations upon the Prophecies of Daniel, and the Apocalypse of St. John,* writes that

> ... giving ear to the Prophets is a fundamental character of the true Church. For God has so ordered the Prophecies, that in the latter days *the wise may understand, but the wicked shall do wickedly, and none of the wicked shall* understand. *Dan.* xii, 8, 10. The authority of Emperors, Kings and Princes, is human. The authority of Councils, Synods, Bishops, and Presbyters, is human. The authority of the Prophets is divine, and comprehends the sum of religion, reckoning *Moses* and the Apostles among the Prophets. . . . Their writings contain the covenant between God and his people, with instructions for keeping this covenant; instances of God's judgements upon them that break it: and predictions of things to come.[21]

Just as Newton (in all likelihood) utilized Bentley's Boyle Lectures in 1692 as a platform to publicize how his "systeme" demonstrated God's general providence, so too he apparently utilized the same platform to demonstrate God's special providence as revealed in lists of fulfilled prophetic predictions in Whiston's 1708 Boyle Lectures. In these lectures, which, as we have seen, were probably suggested by Newton[22] and which were dedicated to the archbishop of Canterbury, Thomas Tenison, Whiston argued that the war with the French represented proof of God's continued special providence for his chosen people, the English Protestants. Their victory would bring the Reformation to its predicted fulfillment while illustrating God's special power and providence in directing the course of world history.

During the first eighty years of the Royal Society, the chief apologetic aim of these scientist-theologians was to balance general and special providence. The Newtonians finally achieved a kind of synthesis of the two. The Newtonian version of the design argument demonstrates a generally provident deity who first creates the universe and the laws of nature and then "miraculously" – that is, continuously – preserves that creation. The argument from fulfilled prophecy discloses, through its lists of prophetically revealed predictions that have come true, a specially provident deity active in governing his creation directly and gives clues for the wise concerning the future course of human and natural history. This argument is particularly emphasized by the Newtonians of the "first" Royal Society[23] and by William Whiston.

Criticisms of the synthesis of general and special providence and Whiston's continued efforts in the course of the century to link them

Many members of the early Royal Society sought to promote a strong conception of generally provident nature and regular natural law while also retaining God's special providence via miracles or prophecies. But with the retirement of Sir Hans Sloane from the post of president and the subsequent election of Martin Folkes in 1741, the apologetic goal of the first eighty years of the society was abandoned. William Stukeley, a close friend of Newton and Whiston, rector of St. George's, Bloomsbury, and Fellow of the Royal Society, recorded the nature of Folkes's palace revolt against the apologetic goals of the "first" society. Folkes, according to Stukeley,

> . . . chuses the Council & Officers out of his junto of Sycophants that meet him every night at Rawthmills coffee house, or that dine with him at the Miter, fleet street. He has a great deal of learning, philosophy, astronomy: but knows nothing of a future state, of the Scriptures, of revelation. He perverted (the) Duke of Montagu, Ld. Pembroke, & very many more of the nobility, who had an opinion of his understanding; & this has done an infinite prejudice to Religion in general, made the nobility throw off the mask & openly deride & discountenance even the appearances of religion, wh has brought us into that deplorable situation we are now in, with thieves, & murderers, perjury, forgery, &c. He thinks there is no difference between us & animals; but what is owing to the structure of our brain, as between man & man. When I lived in Ormond in 1720, he set up an infidel Club at his house on Sunday evenings where Will Jones the mathematician, & others of the heathen stamp, assembled. He invited me earnestly to come thither but I always refused. *From that time he has been propagating the infidel System with great assiduity, & made it even fashionable in the royal Society, so that when any mention is made of Moses, of the deluge, of religion, scriptures, &c., it generally is received with a loud laugh.*[24]

The institutional shift in the "second" society divorcing religion from science seems to have been accomplished within the society itself more through ridicule and mockery than as a result of explicitly argued tracts and pamphlets. The same is the case with the Scriblerian wits who savaged Whiston's antideistic Newtonian doctrines with acid wit.

In 1726, Whiston had a craftsman build scale models of the Tabernacle of Moses and of "the Temple of *Jerusalem,* serving to explain *Solomon's, Zorobabel's, Herod's,* and *Ezekiel's* Temples; and had Lectures upon that at *London, Bristol, Bath* and *Tunbridge Wells.*"[25] Until his death, Whiston labored on using his Newtonian method of scriptural interpretation to reveal to the world the role of current events in leading up to the millennium and the physical reconstruction of the Temple in Jerusalem. In his millennial lectures of 1750 immediately following the first of two mild earthquakes, Whiston refers to the notorious case of Mary Toft, who in 1726 had caused a great stir with her claim to have given birth to at least seventeen baby rabbits (see Figure 4).

Whiston's millennial lectures formed part of his demonstration of the providence of God in human affairs, and by 1729 or 1730 it proved too tempting a target for the Scriblerians, John Gay and Jonathan Swift, to pass over. In 1729 or 1730, Gay composed "A True and Faithful Narrative of What pass'd in London during the general Consternation of all Ranks and Degrees of Mankind."[26] Gay's satire purports to be the report of a Londoner who had paid a shilling to hear Whiston lecture one Tuesday on some subject (never mentioned) only to find that Whiston "thought himself in duty and conscience, oblig'd to change the subject matter of his intended discourse." "Whiston's" lecture was brief and apocalyptic:

> Friends and Fellow-Christians, all speculative Science is at an end; the Period of all things is at Hand; on *Friday* next this World shall be no more. Put not your confidence in me, Brethren, for tomorrow Morning [Wednesday] five Minutes after Five the Truth will be Evident; in that instant the Comet shall appear, of which I have heretofore warn'd you. As ye have heard, believe. Go hence, and prepare your Wives, your Families and Friends, for the universal Change.[27]

There were two skeptics in the crowd, and Whiston gladly refunded their lecture fee, but the rest spread the news of imminent catastrophe, and soon it "was the subject-matter of all conversation." Naturally, nobody believed this prophetic prediction at first. Stockbrokers thought it a ploy by the Court to depress stock prices so "that some choice favourites might purchase at a lower Rate; for the *South-Sea,* that very Evening fell five *per Cent,* the *India* eleven." A Quaker, Zachary Bowen, told Whiston's disciple that "thy tidings are Impossibilities, for were these things to happen, they must have been foreseen by some of our Bretheren. This indeed (as in all other spiritual Cases with their set of People) was his only reason against believing me."[28]

But the skeptics were soon convinced when Whiston's predicted comet arrived on the next day, Wednesday, nearly on time (it was two minutes early). Apparently the world was doomed to end in two days. At this point, Gay's satire takes wing in earnest. He gleefully describes the reaction of various individuals to the new prophecy of doom. His strokes are hilarious: 123 clergymen are ferried to Lambeth to petition the archbishop of Canterbury to provide a short prayer for their end-of-the-world liturgy, "there being none in the Service upon that occasion";[29] the director of the Bank of England, Sir Gilbert Heathcote, required all the fire stations "to have a particular Eye upon the Bank of England. Let it be recorded to his Praise that in the general hurry, this struck him as his nearest and tenderest concern";[30] 7,245 men flocked to the churches to marry their mistresses;[31] and "at St. *Bride's* Church in *Fleetstreet,* Mr. *Woolston* (who writ against the Miracles of our Savior) in the utmost Terrors of Conscience, made a publick Recantation. Dr. *Mandevil,* (who had been groundlessly reported formerly to have done the same) did it now in good earnest at St. *James'* Gate; as did also at the Temple Church several Gentlemen who frequent the *Coffee-Houses* near the Bar";[32] finally, women confessed to their forgiving husbands the true identity of the fathers of "their" children.[33]

This gentle mocking takes on a sharper tone when Gay describes what happened when doomsday came and went:

> The subject of all Wit and Conversation was to ridicule the prophecy, and railly [*sic*] each other. All the quality and gentry were perfectly asham'd, nay, some utterly disown'd that they had manifested any Signs of Religion.
>
> But the next day, even the Common People, as well as their Betters, appear'd in their usual state of Indifference. They Drank, they Whor'd, they Swore, they Ly'd, they Cheat'd, they Quarrell'd, they Murder'd. In short, the world went on in the old channel . . . Mr. Woolston advertis'd in that very *Saturday's Evening Post* a new Treatise against the Miracles of our Saviour.[34]

Most ironically, the 1750 London earthquakes (discussed in the next section of this chapter) occasioned precisely such "consternation" and afflicted "all Ranks and Degrees of Mankind" nearly in the manner described here by Gay about 1730 in the context of an imaginary lecture by Whiston.

It is difficult to tell if the target of Gay's satire is really Whiston or merely human nature in general. It is probably a bit of both. Whiston's 1726 lectures and his theory of comets were almost certainly the proximate cause of Gay's witty remarks about the fickleness of humanity.

Whiston, of course, does not see himself in this light. Whiston never considers himself to be a divinely inspired prophet. He sees himself as a scientific interpreter of the providential nature of the two books of God, nature and Scripture. In his pamphlet describing the "meteor" (that is, the northern lights) observed in 1716, he acknowledges that as yet there is no adequate scientific or religious explanation for this phenomenon and emphasizes that not every unusual event can be taken for "Vulgar Prognostications and Omens" about the future, as he claims is possible in the case of the 1750 earthquakes:

> And the Reason is plain; that the Author of Nature has in one Case declar'd his Meaning as to those Events; but has not done so in the other: Which Meaning, without such Declaration, it is great Folly and Presumption for poor Mortals to pretend to discover. God may, no doubt, if he pleases, make use of his Original Settlement of Natural Causes, without recourse to his own immediate Power; not only to foreshew and foretel, but also to execute his own Purposes. A Deluge or Conflagration, as I have elsewhere shewn, may be brought about by a Comet, in its Descent or Ascent, without the Introduction of any thing strictly Supernatural or Miraculous. And the Reason is plain, that Nature is only God's appointed Order for his own Creatures Operations, by Powers deriv'd from himself; and therefore, He that foreknew all things at first, could accordingly fore ordain, prepare, and pre-dispose any Parts of his own System, not only to foreshew but really to bring about what Acts of Mercy or Judgement he, in his Divine Wisdom, shall think fit for his Creatures. This World is *God's* World: and what Nature does, is in reality and ultimately done by the God of Nature. . . . On which account, in a sober Sense, all the Phenomena of the World are deriv'd from a Supernatural and Divine Power. But, then, till God's particular Meaning be discover'd to us, this will not enable us to foretel future Events from them. I am indeed under a peculiar Temptation my self to Wish and Suppose that this and the like unusual Appearances, may be Prognosticks and Forewarnings of the Coming of those Great Concussions and Mutations which I expect soon in the World, to the Depression of Antichristianism and the Revival of true Christianity in its stead. But this notwithstanding, I shall keep to Truth and Evidence; and till I see apparent Marks of a Supernatural Intention herein, shall not pretend to interpret the Secrets of Divine Providence in favour of any Opinions, I have entertain'd on other Foundations.[35]

All of Whiston's Newtonian nuances are included in this text: the tension between general (natural) providence and special (supernatural, miraculous) providence; the possibility of collating the word of God revealed in the twin books of nature and Scripture when "scientifically interpreted"; the difficulty of this project and the care with which it must be prosecuted; the necessity of adhering to "genuine" evidence rather than to vain hypotheses.

For Gay's satiric purposes, however, it was best to treat Whiston as an obsessed prophet rather than as a serious and sober Newtonian biblical interpreter. Even Pope, who learned so much from Whiston's scientific lectures at Button's Coffeehouse, suggests that Whiston and his religious theories are yet another judgment on a sinful nation, along with the Great Fire. After playfully cataloging the excesses of Restoration England (especially "the Abomination of Playhouses"), Pope notes England's eventual deliverance and subsequent backsliding: "But when whoring and Popery were driven hence by the Happy *Revolution;* still the Nation so greatly offended, that *Socianism, Arianism,* and *Whistonism* triumph'd in our Streets, and were in a manner become Universal."[36]

Swift, too, we have noted, satirized Whiston's Arianism and his project for determining the longitude (see Chapters 1, 4). But Swift directly mocks the whole tradition of scientific biblical interpretation, in *Dean Swift's True, Genuine, and Authentic Copy of that most Strange, Wonderful, and Surprizing Prophecy Written by St. Patrick, the Patron of Ireland, Above a Thousand Years ago: Faithfully Translated from the Irish Original above two hundred years since, in the reign of K. Henry VII. New Publish'd with Explanatory Notes* (1740). Directed primarily against Bishop Thomas Sherlock's *Use and Intent of Prophecy,* Swift cuts at the pridefulness of biblical interpreters who so effortlessly reveal to us God's providential involvement in human affairs. Swift presents an extremely opaque poem as a prophecy uttered by St. Patrick and interprets it to refer to Marlborough's defeat of the French. At the conclusion, Swift modestly proposes that "some of these Predictions are already fulfilled; and it is highly probable the rest may be in due Time: And I think, I have not forced the Words by my Explication into any other Sense than what they will naturally bear."[37]

The Scriblerians' ridicule of biblical interpretation in general and Whiston's millennial interpretation in particular, as well as the idle scoffing on the part of the wits within the Royal Society of Martin Folkes's era, were not based on coherent philosophical arguments. Through caustic and often sharply pointed ridicule, these wits attacked Whiston's Newtonian attempt to balance God's general and special providence.

The 1750 London earthquakes: Whiston's last attempt to unite God's special providence and his general providence; its reception

On February 8, 1750, a slight earth tremor in London caused minor property damage and mild curiosity about the physical cause. Precisely one month later, on March 8, a stronger tremor caused more serious property damage (though no deaths) and precipitated great concern among the populace about what had caused the quakes, why they were spaced at such a seemingly precise interval, and, especially, about what was going to happen next.[38] Those two living relics of the early era of the Royal Society, Stukeley and Whiston, joined other eminent divines such as Bishop Sherlock of London and Bishop Secker of Oxford (later archbishop of Canterbury) in producing a shower of sermons and exhortations to answer these questions. These attempts to explain the earthquakes as instances of God's specially provident intervention into his generally provident machine of nature underscore the waning vitality of this Newtonian form of antideism.

Stukeley notes that "when so great and unusual a *phaenomenon,* as an earthquake, and that repeated, happens among us; it will naturally excite a serious reflexion in every one that is capable of thinking. And we cannot help considering it, both in a philosophical, and a religious view."[39] Stukeley adduces (in lectures before the Royal Society printed in the *Philosophical Transactions*) the "philosophical" hypothesis that earthquakes result in the purely physical realm from a "non-electric cloud" discharging "its contents upon any part of the earth, when in a high electrified state."[40] Whatever this might mean (there seems to have been some confusion among the Fellows present at Stukeley's lectures), he attributes his hypothesis to "a very curious discourse, from Mr. Franklin of Philadelphia, concerning thundergusts, lightning, the northern lights, and like meteors."[41]

Stukeley is much more clear in his "religious view" of these earthquakes, and from his pulpit in St. George's, Queen Square, he announces that it is God who is throwing the switches because he is "wroth" (Ps. 18:7). God is angry at the sins of Londoners, their riches, pride, luxuries, vanities, pleasures, profaneness, gaming, immoralities, infidelities, and "especially ... the notorious crime of sabbath-breaking, which is the foundation of all, and comprehends all others."[42]

Stukeley, like John Keill, is one of those few Newtonians who do not flinch from the idea of a miraculous break in the natural order of things. It is a sign of God's specially provident presence that he shakes the city sufficiently to remind people of their moral corruption and their need of repentance but not so hard as to produce any serious loss of life – at least not yet. Stukeley writes, "We know the nature of the building of *London* houses; which sometimes fall of themselves, without shaking. Wonderful then is to thought, and a miracle indeed, that every house in this vast city, should twice be agitated, and rocked to and fro; and not one fall, nor one person receive any damage. In vain will the philosophers seek for a solution of this problem, in natural causes only."[43]

The precisely designed and miraculous effect of the earthquakes is meant to shake us out of our complacency and sin, to get us back into church, and to cause us to mend our evil ways. As for what will happen next, Stukeley acknowledges that he is not a prophet: " 'Tis true, we have hitherto escaped. But can we tell how soon God shall let loose the avenging power of another; which may come, for ought we know, while we are speaking of it. And if it must come, happy may it be for us, that it finds us in this place, and so doing."[44]

Whiston, predictably, preferred to use the argument from prophecy to demonstrate God's special providence. On March 6, 8, and 10 he gave three lectures in London in which he situated the earthquakes in the context of ninety-nine signals or prophetic predictions regarding the restoration of the Jews to Jerusalem and the subsequent arrival of the millennium. Prediction no. 7, drawn on the basis of his interpretation of Isa. 5:18–20 and (the apocryphal) 2 Esd. 9:3, indicates that "many and terrible *Earthquakes* are to come upon Mankind, either from the Air above, or the Ground below, or both together."[45] The completion of this prediction is revealed in Whiston's history of the terrific increase in earthquake activity in recent times.[46]

Prediction no. 92 is from Rev. 11:13, which predicts a great earthquake wherein the "tenth part" of a large city will fall and "seven thousand Names of Men" are to be slain. Of course, this prediction has not yet been completed, but Whiston has "good reasons" for believing it refers to London: he cites the recent increase in the number of earthquakes revealed in a historical survey that he has made (especially the two most recent ones), and the "horrid Wickedness of the Present Age."[47] The whole third lecture is devoted to this latter topic and reviles Jews, Catholics, and especially London Protestants for their many and varied sins. The attendance at these lectures was twenty-seven, sixty-five, and forty-three, respectively. There were even reports circulating that Newton had himself scientifically predicted the earthquakes on the basis of the close approach of Jupiter. Whiston blasts such ignorance in the *Daily Advertiser* of March 14, 1750:

> Mr. *Whiston* gives notice, That though he expects many more Earthquakes in the World, within a Year, or two at the farthest, before the Restoration of the *Jews,* as Signals of its Approach, and of the horrible Miseries the wicked Part of the *Jews* and *Christians* will be subject to, while the really Pious and Good will be providentially delivered from

> them, yet does he not in the least believe that Sir *Isaac Newton* foretold any Earthquake; and is sure that *Jupiter,* at the Beginning of this Year 1750, was, and is 400,000,000 Miles off the Earth, and so could not possibly have any Influence on Earthquakes here below.[48]

The bishop of Oxford, Thomas Secker, also cites Rev. 11:13 and connects the predicted devastating earthquake with the two already accomplished: "We are authorised to conclude, that such Visitations may be Tokens of wonderful Changes, in the natural, in the civil World; as indeed the Book of *Revelation* affords Ground to believe, that, sooner or later, they will."[49] Secker does not claim to be prophet enough to determine exactly when the crushing blow will fall, but fall it almost certainly will. A wise man will make a version of Pascal's wager: "Possibly we may feel no more of these Shocks. Possibly no Damage may attend them, if we do. But is Possibility, is Probability, in a Matter so totally out of Sight, firm Ground enough for you to risque your Soules upon?"[50]

In the midst of all these worthies pointing to God's providential and probably continuing involvement with London's fault zones, there arose the "Military Prophet," a lifeguardsman named Mitchell, who predicted that the expected third quake would fall exactly one month after the second quake, on April 5. As April 5 approached, the people of London became panic-stricken. "Publicus," writing in the *General Evening Post,* estimates that over one hundred thousand people left their houses to take refuge in Hyde Park on the night of Wednesday, April 4.[51] The more reliable Horace Walpole reports on April 4 that "this frantic fear prevails so much that within these three days 730 coaches have been counted passing Hyde Park Corner with whole parties removing into the country."[52]

Many Fellows of the Royal Society in 1750 were also clergymen, and many shared Stukeley's and Whiston's opinion that God was speaking to his sinful children (in a generally or specially provident way) through these earthquakes.[53] But these men, like Stukeley and Whiston, published their views on the "religious aspect" of the earthquakes in sermons. The "official" view of the Royal Society was by now expressed solely in their *Philosophical Transactions,* and within these pages even such writers as Stukeley, who were concerned with both the "philosophical" and the "religious" aspect of the earthquakes, confined themselves to the former. As we have seen, Stukeley records, perhaps somewhat bitterly, that since 1720, when Martin Folkes began to exert influence within the society, the prevailing attitude had been "that when any mention is made of Moses, of the Deluge, of religion, scripture, &c., it generally is received with a loud laugh."

Stephen Hales, curate of Teddington, a friend of Stukeley's, and one of the most distinguished scientists of the Royal Society after Newton, addresses himself in his paper to the society solely to the "philosophical" aspect of earthquakes, while recommending the bishop of London's tract on the "religious" aspect of these events.[54]

After 1741, Folkes and his "junto of Sycophants" assumed leadership of the society. While by no means all the members were scoffing deists, no longer did devout members identify the purposes of the group with the vindication of religion, natural and revealed, as in Newton's era. This change of direction was most completely articulated by Hume, who writes to his physician friend Dr. John Clephane, regarding the two earthquakes,

> I think the parsons have lately used the physicians very ill, for, in all the common terrors of mankind, you used commonly both to come in for a share of the profit: but in this new fear of earthquakes, they have left you out entirely, and have pretended alone to give prescriptions to the multitude. I remember, indeed, Mr. Addison talks of a quack that advertised pills for an earthquake, at a time when people lay under such terrors as they do at present. But I know not if any of the faculty have imitated him at this Time. I see only a Pastoral Letter of the Bishop of London, where, indeed, he recommends certain pills, such as fasting, prayer, repentance, mortification, and other drugs, which are entirely to come from his own shop. And I think this is very unfair in him and you have great reason to be offended; for why might he not have added, that medicinal powders and potions would also have done service: The worst is, that you dare not revenge yourself in kind, by advising your patients to have nothing to do with the parson; for you are sure he has a faster hold of them than you, and you may yourself be discharged on such an advice.
>
> You'll scarcely believe what I am going to tell you; but it is literally true. Millar had printed off some Months ago a new edition of certain philosophical Essays [i.e., the second edition of the *Philosophical Essays Concerning Human Understanding*] but he tells me very gravely, that he has delay'd publishing because of the Earthquakes. I wish you may not also be a Loser by the same common Calamity. For I am told the Ladies were so frightened that they took the Rattling of every Coach for an Earthquake, & therefore wou'd employ no Physicians but from amongst the Infantry: Insomuch that some of you Charioteers had not gain'd enough to pay expense of your Vehicle. But this may only be Waggery & Banter, which I abhor.[55]

Hume is not the only one who was unable to take this spectacle as witness of God's providence. Horace Walpole also provides a scathing attack on the old Newtonian synthesis of general and special providence. It was just because Whiston was serious in holding these views that Walpole believed him to be crazy:

> When Whitfield preaches, and when Whiston writes,
> All cry, that madness dictates either's flights.
> When Sherlock writes, or canting Secker preaches,
> All think good sense inspires what either teaches.
> Why, when all four for the same Gospel fight,
> Should two be crazy, two be in the right:
> Plain is the reason; ev'ry son of Eve
> Thinks the two madmen, what they teach believe.[56]

When the predicted earthquake failed to materialize on April 5, a "Gentleman in Town" (probably Paul Whitehead), produced a brief pamphlet, written as though the earthquake had really happened, in which he gave an "Exact List" of the people found in the debris. Among the victims, he says, was the bishop of London, Thomas Sherlock:

> The very first Man was sunk by this Earthquake, was the B——— of L———: It seems he might have excaped, but his Zeal was so great in distributing Copies of his Letter, which, good Man, as the time drew near, he gave away in Bundles, Thirteen to the Dozen, to any Body that would accept of them, that he took no Manner of Heed to his Steps, and so entirely lost himself.[57]

"Old Whiston" escaped because he, "upon the first Beginning of the Trembling, set out on Foot for *Dover,* on his way to *Jerusalem,* where he has made an Appointment to meet the Millennium: It is thought, if he makes tolerable Haste, he will arrive first."[58] The "Military Prophet," Mitchell, who had made the April 5 prophecy in the first place, produced a pamphlet explaining that the earthquake did not happen as predicted because so many sinners had heeded his words and tried to escape destruction in Hyde Park, Lincoln's Inn fields, and the country; but, he asserted, it was still coming.[59] Finally, he was locked away in a madhouse.

II. Newton's deism and Whiston's antideistic criticism

In Section I we witnessed how Newton, earlier in the eighteenth century, and Whiston, until his death in 1752, attempted to preserve both the creative, generally provident God of Genesis and the specially provident God described throughout Scripture who is sufficiently powerful to intervene in nature through miracles but also to make special interventions that fulfill biblical prophecies. Whiston and Newton reject any deistic interpretation of the Bible that aims at reducing God's providential power as it is revealed in the Bible.

The provident God of Christian revelation – a creating, preserving, miracle-working, prophecy-giving and prophecy-fulfilling God – remains the fixed point of reference in their methodological disputes either with the deists or among themselves. Voltaire's summation of Newton's antideistic view about the *nature* of God as revealed in the Bible is accurate: "Sir Isaac Newton was firmly persuaded of the Existence of a God; by which he understood not only an infinite, omnipotent, and creating being, but moreover a Master who has made a Relation between himself and his Creatures."[60]

Bentley's, Whiston's, and Newton's own versions of the design argument demonstrate the creative God of general providence. Newton's and Whiston's work on literally – that is, singly determinate – fulfilled historical prophecies accommodates the idea of the generally provident God of natural religion and illustrates the revealed God of Christianity, a God who is Lord God and who intervenes directly in creation through his acts of special providence. Newton's "original suggestion" that Whiston sermonize on this latter topic in his Boyle Lectures is especially significant evidence that Newton opposed any deistic attempt to belittle or ridicule properly – that is, "literally" – interpreted historical fulfillments of biblical prophecy. The concept of a specially provident God, as well as an indivisible monotheistic one, was, they believed, clearly grounded in Scripture. As Whiston states, "modern unbelievers" or deists have disputed "every Inch of Ground where a Providence and the Power of God us'd to be suppos'd in the World. The Almighty has at last been pleas'd, by the noble Discoveries of late afforded us, to put an end to that question."[61]

But not all of the Bible reveals the provident nature of the deity. Much of the Bible, in Newton's and Whiston's view, is reducible to plain history. And it is Newton's approach to these narrative historical portions of Scripture that makes him a kind of deist and causes Whiston to attack him. For, as Westfall has pointed out, Newton's attempts at demystification led him to equate these historical portions of the biblical text with other ancient historical sources and often to correct the former in terms of the latter. Richard S. Westfall's analysis of Newton's manuscript "Theologiae Gentilis Origines Philosophicae," which dates from "soon after 1683,"[62] makes three major points. First, Westfall demonstrates the centrality of this work by showing how Newton continued to make veiled references to it throughout his life: in revisions to the *Principia* in the 1690s, in what becomes Query 31 of the *Optics* ten years later, and in the "General Scholium" of the 1713 edition of the *Principia*. This manuscript, Westfall asserts, formed the basis for Newton's *Chronology of Ancient Kingdoms, Amended* (1728) as well. Westfall rightly terms it "the most important theological work he ever produced."[63]

Next Westfall analyzes the content of this manuscript and shows us a critically minded historical chronologist who judges various ancient pagan sources to be more authoritative than selected portions of narrative scriptural history. It is simply not the case, in the light of this key manuscript, that throughout Newton's historical and theological writing "Biblical chronology remained the touchstone, incontrovertible, by which all heathen chronologies were tested," as Manuel asserts.[64] In writing history, Newton, Westfall argues, weighed many kinds of literary evidence and, as his "Theologiae Gentilis" shows, for Newton pagan sources constituted evidence at least equal in authority to Scripture and occasionally exceeding the authority of the Bible.[65]

Unlike such seventeenth-century polymaths as Gerard Vossius, Samuel Bochart, and Sir John Marsham, all of whom agreed that the Scriptures are the certain criterion by which all other ancient traditions must be judged, Newton questions Moses' historical account of the era around the time of the Flood and revises it in the light of other pagan literary sources. For example, Newton revises Moses' sketchy account of the founding of the Assyrian empire by Nimrod, son of Cush, by relying upon such ancient sources as Berossus the Babylonian and Sanchuniathon the Phoenician (as preserved in Eusebius and Josephus and given currency by the seventeenth-century polymaths).[66]

Finally, Westfall's analysis of Newton's manuscript reveals a truly radical figure who sifts all the ancient sources and discovers evidence of an uncorrupted, "original true Religion" preserved in part by the Gentiles and in part by the Hebrews. This discovery convinces Westfall of "the ultimate heterodoxy both of the treatise and of its author."[67] For Newton, this most ancient and true religion focused ritualistically upon the worship of perpetual and sacred fires, or *prytanea.* The heliocentric universe itself is the original type for all the prytanea found in the religious observances of the Etruscans, the most ancient Greek cities, the Persians, the Egyptians, and the Hebrews, whose Temple, with its central altar fire, was modeled on this most ancient religion. Newton's enduring interest in the Jewish Temple is partly millennial and partly because it is a type of "the true Temple of ye great God" Nature.[68] Noah carried the "original true religion" to Egypt, where it was increasingly corrupted by an unsavory mixture of human credulity and superstition, such as the worship of stars. During their Egyptian captivity, later generations of Jews rediscovered and restored for a time the basic elements of the *prisca theologiae,* but they, too, fell away, despite the pleadings of their prophets. For Newton, Jesus was just another prophet sent to restore lapsed humanity to the true religion.

The first two doctrinal precepts of this ancient tradition, according to Newton, were love of God and love of neighbor, precepts most deists thought a part of natural religion and therefore true. The third was that the providential power of God had revealed to mankind the coming of Jesus in fulfilled historical prophecies. This last point, of course, constitutes the heart of Whiston's antideism and, in its denial of consubstantiality with the Father to Jesus, his antitrinitarianism.

The methodology by which Newton arrives at this heterodox theology is genuinely deistic in the manuscript. His assumption that the Bible was one among many equally authoritative historical sources was closer in spirit to the attitude toward Scripture ascribed to the circle who met at Martin Folkes's house (except that Newton lacked their mocking spirit and held the prophetic part of Scripture to be sacred) than it is to the traditionalist attitude that every word of Scripture is derived from God and in some sense true.

Martin Folkes helped Thomas Pellet edit Newton's *Chronology of Ancient Kingdoms, Amended* after Newton's death so that John Conduitt, Newton's nephew, could publish it. This work is a sanitized and truncated version of Newton's more radical manuscript.[69] Since Newton typically kept this aspect of his heterodoxy and the manuscript containing it secret during his lifetime, there is no evidence that his manuscript was ever directly attacked. However, even the heavily edited published versions of Newton's theological works, including his *Observations on the Prophecies of Daniel and the Apocalypse of St. John* (1733), provoked angry responses from more orthodox Anglicans frightened by what they perceived to be Newton's antiscriptural bias. Newton's tendency to bend the Bible to conform with other ancient traditions prompted Arthur Young to write that he was "sorry to see Principles so favouring the Schemes of the Deists, with so great a Name affix'd to them."[70] Even Whiston, who never shrank from pointing out scriptural (and doctrinal) corruptions to orthodox Anglicans on the strength of his Newtonian method of interpretation, broke with Newton on Newton's *Chronology*.

Whiston, after Newton's death, finally stated in public that Newton also held the antitrinitarian views for the sake of which Whiston had sacrificed his career:

> Sir I. N. was one who had thoroughly examined the State of the Church in its most critical Juncture, the fourth Century. He had early and thoroughly discovered that the Old Christian Faith, concerning the Trinity in particular, was then changed; that what has been long called

> *Arianism* is no other than the Old uncorrupt *Christianity;* and that *Athanasius* was the grand and very wicked Instrument of that Change. This was occasionally known to those few who were intimate with him all along; from whom, notwithstanding his prodigiously fearful, cautious, and suspicious Temper, he could not always conceal so important a Discovery.[71]

Whiston goes on in the same vein and urges Thomas Pellet, the executor of Newton's literary estate, to publish all of Newton's theological papers and not to suppress them for "any Ecclesiastical or Political Reasons":

> I extend this reasoning to Dr. *Pellet* in particular, to whom Sir I. N.'s Papers were Legally committed, to peruse and select what were fit for publication; and who, as I am informed, gives out that Sir I. N. did not at last believe the Bible or the Christian Religion. And I conjure him, on the Peril of his being otherwise charged with open Falsehood and Forgery, if he continue that Report, to take care that those original Papers from which he pretends to collect that Imputation, be ready to be produc'd under Sir I. N.'s own Hand for his justification.[72]

Whiston's charge to Pellet that he publish Newton's antitrinitarian papers comes, ironically, at the conclusion of his 120-page attack on Newton's *Chronology*. Whiston detected in this work the taint of the deistic methodology rampant in the "Theologiae Gentilis." Whiston knew the general outline of Newton's chronological system prior to its publication in 1728 from conversations with Newton and Roger Cotes. He did not think much of it then,[73] and his "Confutation" shows why.

Newton's basic mistake in his chronological organization of the events of ancient history, according to Whiston, is methodological. Newton runs afoul of Whiston's third postulate of interpretation: "What Ancient Tradition asserts of the constitution of Nature, or of the Origin and Primitive States of the World, is to be allow'd for True, where 'tis fully agreeable to Scripture, Reason, and Philosophy." Instead of correctly using the best copies of scriptural texts and the solid evidence of natural philosophy to interpret other ancient authorities, Newton, Whiston says, provides a sagacious romance based only partly on "Historical Authors; but partly, upon the Poetick Stories of Mythologists laid together by himself; and partly, nay principally, upon fond Notions, Vehement Inclinations, and Hypotheses of his own." The evidence for Newton's chronology, according to Whiston, is not comparable to the sure demonstrations Newton gives in natural philosophy; rather, the whole work is "an Imaginary or Romantick Scheme . . . built upon no manner of real foundation whatsoever."[74]

Newton, Whiston says, ignores the best accounts of ancient chronology – those preserved in the Samaritan Pentateuch, the Septuagint version, and

Josephus – in favor of the more corrupt "English version made from the *Masorete Hebrew*. [the worst copy of all.]" (brackets in the original). The Masoretic text, which Whiston does not hesitate to recommend when it agrees with his theories (see Chapter 3), has many fewer chronological references than the other versions. Newton's reliance upon it was "his *fundamental Mistake*, and from which almost all the rest of the Mistakes in his Chronology do naturally and sometimes unavoidably flow." Because of this fundamental methodological error, Newton was obliged to subscribe to the view that the identity of the Egyptian King Sesostris (from Josephus out of Manetho) is the same as the biblical Sesac. And the "natural Consequences" of this identification, which is also held by the seventeenth-century chronologist John Marsham (who also uses the figures given in the Masoretic text), are to remove "the Argonautick Expedition, the Exploit at *Thebes*, the Trojan War, with the return of the *Heraclides* into *Peloponenses*, (all of which Antiquity placed a great deal later than *Sesostris*) much lower than their proper Evidence could possibly allow."[75] Once this fundamental error is corrected, that is, once one rejects the "*Masorete* numbers," and "*Marsham's* Hypothesis" that Sesostris is Sesac, Newton's entire scheme of chronology – which brings events recorded in Gentile history forward in time, making them generally postdate the reign of Solomon – collapses.[76]

Whiston believed that Newton's most revolutionary contribution to the problem of synchronizing sacred and profane history was his application of the precession of the equinoxes to the dating of the Argonautic expedition.[77] But he thought that even this procedure, which accords with Whiston's third postulate in that it attempts to date an ancient event in Gentile hisory by means of "reason and natural philosophy," had been improperly worked out by Newton.

To utilize the equinoctial precession as a means of dating the Argonautic expedition, Whiston argues, requires a reliable ancient record of the location of the sun in relation to the sphere of the fixed stars at the time of the equinox. From the fixed rate of precession, which Newton had calculated in the *Principia*,[78] one could then work backward from the present to date the event exactly. Whiston accepts Newton's calculation of the rate of precession[79] but argues that Newton's interpretation of the rough descriptions given by the fourth-century-B.C. Greek astronomer Eudoxus (preserved in a second-century-B.C. fragment from Hipparchus) of the position of the equinoctial colures (drawn through the equinoctial points) at the time of the Argonauts is ill founded. Newton, Whiston thought, mistakenly interpreted Eudoxus to mean that one of the equinoctial colures passed through the middle of Aries, although Eudoxus specifically states that the equinoctial colure passed over the "back" of Aries. From the known rate of precession, Newton calculated backward from the location of this equinoctial colure in his day to its location in the "middle" of Aries (as he interprets Eudoxus) and so astronomically dated the time of the Argonauts at 939 B.C., well after the time of Solomon. (As Manuel points out, from this point of reference it is easy to date the fall of Troy, reported by Herodotus to have occurred one generation after the Argonauts' voyage, and also the founding of Rome by Aeneas, a refugee from Troy.)[80]

Whiston performs the same calculation, utilizing Eudoxus's description of the location of the equinoctial colure at the "back" of Aries (which makes a difference of 5¾ degrees from Newton's location of it in the "middle" of Aries). When Whiston calculates backward from the position of the colures in his day, he arrives at a date of 1353 B.C. for the voyage of the Argonauts: "So that this Place of the *Colures* at the Argonautick Expedition is so far from removing it above 300 Years lower, as Sir I. N., would fain have it; that it obliges us to advance it near 100 Years higher than our modern Chronologers have placed it. Nor can this Consequence be avoided by any Subtility or Art whatsoever."[81]

Manuel terms Whiston's analysis of Newton's dating of the voyage of the Argonauts as both "destructive and convincing," and "fair and acute."[82] But though Newton may have made mistakes in its application, this procedure, Whiston felt, was fundamentally sound. Far more damaging in Whiston's eyes was Newton's synchronization of Gentile and Hebrew history on the basis of the Masoretic text rather than the more accurate older versions. Whiston claims that Newton had thereby produced a curiously "novel system, which subverts almost all the most ancient Chronology we have hitherto had; and bids fair, if it prevail, for the introduction of the utmost Scepticism into all ancient History whatsoever."[83]

The radical deism of the manuscript on the origins of Gentile theology is missing from Newton's *Chronology,* but traces of such an attitude were detectable to that arch-enemy of deism, William Whiston. Acceptance of dating schemes based on corrupted texts subverts biblical authority, however slightly, although Whiston willingly acknowledges that this was not Newton's intention and repeatedly mentions "Sir I. N.'s firm and unshaken Belief of the Sacred History."[84]

To summarize, Newton and Whiston were antideistic in their views about the existence of a generally provident creator, a Lord God who is also willing to intervene in his creation through acts of special providence such as miracles and fulfillments of prophecies. Newton's manuscript, however, shows that he is much closer to the spirit of deism than Whiston in his willingness to categorize the Bible as one more piece of literary evidence from the past. His *Chronology of Ancient Kingdoms, Amended* hints at this position, as Newton attempts to synthesize evidence from a variety of literary and astronomical sources with a corrupt biblical text that led him to a radical and incorrect ancient chronology. Whiston perceives this deistic tendency and attacks it. In his attack on Newton's *Chronology,* Whiston is far more successful in attacking the accuracy of Newton's dating than he is in defending the Newtonian synthesis of divine general and special providence.

III. David Hume's religious skepticism and the breakdown of the Newtonian synthesis of general and special providence

Whiston believed that Newton's approach to Scripture in his *Chronology* helped the deist cause by undermining the authority of that part of Scripture relating to ancient Gentile history. But Whiston was adamant that Newton remained essentially pro-Scripture and antideistic in his interpretation of those parts of the Bible in which God reveals his general and special providence. The generally provident creation of the earth that Moses described as if he had been present, with its beautifully designed mechanism of nature and natural laws, together with properly interpreted biblical reports of accomplished miracles and fulfilled prophecies, show the true nature of the one Lord God. Whiston is in earnest when he challenges Pellet to produce the documentary evidence for Pellet's accusation that Newton "did not at last believe the Bible or the Christian Religion."

But once one admits that the Scripture is in part a human and in part a divine revelation, the problem of determining which part of Scripture is which becomes paramount. Whiston believed that his Newtonian method of interpreting the singly determinate, historical meaning of prophetic history (and, not incidentally, discovering in the process false, anti-Christian, human corruptions) was analogous to the experimental method of the Royal Society or the commonsense method of weighing evidence found in courts of law. Whiston and Newton both began their interpretation of Scripture from the Christian perspective, according to which the greater part of the Bible is divinely inspired. There are other initial starting points that yield radically different conclusions, as noted in the case of Anthony Collins, who, as we have seen, charged that Whiston's Newtonian method merely produced a "Whistonian Bible" (see Chapter 3, Section II).

One of the most radically different approaches to the Bible in the eighteenth century was adopted by the Scottish *philosophe* David Hume. An analysis of Hume's religious skepticism is instructive, because Hume reveals most clearly the danger to theism when theists attempt to ground their theism in rational demonstrations. Hume simply begins by assuming, in contrast to Whiston and Newton, that all of Scripture is a human production, that no part of it is divinely revealed. This starting point inevitably follows the introduction of doubt regarding portions of Scripture and the proper method for sifting the human part from the divine part. From this initial assumption, Hume proceeds to destroy the Newtonian attempt to synthesize general and special providence.

In his "Of a Particular Providence and a Future State" (as well as in his more leisurely attack in the *Dialogues*), Hume undermines the design argument, denying that the existence of a generally provident God can be certainly inferred as the cause of the orderliness of the world of nature. In the essay "Of Miracles," Hume also attacks the idea of a specially provident God who directly intervenes in the regular course of nature to create miracles, which Hume defines as violations of natural law. Hume explicitly widens his attack to include prophecies.

General providence and Hume's "Of a Particular Providence and a Future State"

The essay "Of a Particular Providence and a Future State" is designed to show that the inference to a generally provident creator-designer from the "order" exhibited in the phenomena of nature is only speculative and no more certain than any other hypothetical account of natural order. Hume means by "particular providence" in the title the specific and unique act of the creator in designing the apparent order observed in nature. He thus uses the phrase in the sense of "general providence," and not in the sense reserved for specific (or particular) acts of nature transcending special providence.

Hume argues in the second part of the essay that the denial of the general providence of God as the cause of nature has no effect on the conduct of society.[85] Whether the uneducated masses needed an establishment religion to preserve the order of society was a much disputed question in the Enlightenment period, and Hume was anxious to refute the claim of the necessity of establishment religion as a social policeman of the kind apparently envisioned by Stukeley, who blamed murder and thievery on the "discountenancing" of religion.

But Hume's essay goes farther and undermines the basic causal inference from the "order of work" to the "forethought of the workman." In reasoning from effects to causes, Hume argues that the inferred cause must be proportioned precisely to the observed effects.[86] But the "religious hypothesis" of a generally provident creator-designer is only one possible way among many for explaining observed phenomena, both physical and moral. If we insist that the cause of nature be suited precisely to the effects found in nature, then the only possible rational result is confusion and doubt.[87] By all means, Hume counsels, draw whichever of the possible inferences about the cause of nature you wish, but recognize its dubious nature and allow freedom to others to draw different inferences, especially because such speculations have no effect on the morals of society. "Here," says Hume, "you ought to rest."[88]

In casting doubt on the use of the design argument to infer a generally provident creator-designer, Hume attacks one of the fundamental arguments of the "first" Royal Society. Even more important, however, is the relationship between this familiar argument against generally provident design and the argument in the essay "Of Miracles" that attacks the corollary concept of a specially provident God who directly disrupts the regular course of nature through miracles.

The attack on special providence in "Of Miracles"

As in his essay on generally provident design, so too in the case of miracles Hume believed that "a wise man proportions his belief to the evidence." The evidence for miracles is incompatible with the laws of nature as established by experience.[89]

In the first section, Hume concentrates on the conflict between each individual's experience of the laws of nature and a miracle. Although Hume considers causation rationally indemonstrable, he does recognize a regularity of succession in practical human experience. Laws of nature, for Hume, are established because of universally firm and unalterable human experience of such a succession of events. A miracle, however, is by definition a "violation of the laws of nature."[90] Because a miracle is a violation of that for which we have "a firm and unalterable" experience, the evidence for a miraculous event must be of a degree of strength that is impossible to obtain. The more miraculous an event appears to be, the more contrary to the normal sequential experience of natural law and, consequently, the less believable it is.

Hume acknowledges the possibility of apparent exceptions to natural law but argues, with the example of the subcontinental Indian's ignorance of the freezing of water in cold climates, that such exceptions are only apparent exceptions and are due to limited experience. They are not really violations of the law of nature but result from a lack of experience of the regularity with which water freezes below a certain temperature.

To believe in a miracle requires evidence that is impossible to obtain because it runs counter to our unalterable experience to the contrary. And when one considers how open to deception and error are the historical testimonies of such "prodigies," the "plain consequence" is "that no testimony is sufficient to establish a miracle, unless the testimony be of such a kind, that its falsehood would be more miraculous, than the fact, which it endeavours to establish."[91]

In the second section of the essay, Hume examines the sorts of testimony upon which the credibility of miracles is founded to see if there is any type that satisfies the criterion that the falsehood of the testimony would be a greater miracle than the miracle that the testimony purports to establish in the first place. Hume finds no examples where the falsehood of such testimony would be more miraculous than the event it is used to confirm, especially because none of the testimony originates with hardheaded, enlightened Scotsmen. In the penultimate paragraph, Hume concludes with his famous direct reference to the miracle stories of the Bible:

> I desire any one to lay his hand upon his heart, and after a serious consideration declare, whether he thinks that the falsehood of such a book, supported by such testimony, would be more extraordinary and miraculous than all the miracles it relates; which, is, however, necessary to the measure of probability above established.[92]

What has heretofore been largely neglected in the voluminous discussions of the essay "Of Miracles" is the last paragraph. In this paragraph, Hume drops a bombshell that exposes the futility of the attempt of the Newtonians to retain the biblical God of special providence by using the argument from fulfilled prophecies. Hume asserts, "What we have said of miracles may be applied, without any variation, to prophecies; and indeed, all prophecies are real miracles, and as such only, can be admitted as proofs of any revelation."[93]

To understand what Hume means by saying that all prophecies – that is, all prophecies capable of buttressing revealed religion – are genuine miracles, some distinctions must be made among the various kinds or classes of prophecies. In the first place, for Hume and other eighteenth-century writers, as we have seen, "prophecy" is synonymous with "prediction."[94] Beyond that, there are at least three important classes of prophetic predictions.

The first class of predictions is based on the rational examination of empirical data that may be verified by any examiner. These are the familiar predictions of scientists. These scientific prophecies are always subject to testing or experiment to verify their accuracy. For example, in 1823, Bessell prophesied, or, as we should say today, postulated, the existence of the planet Neptune as the result of observing perturbation in the motion of the planet Uranus; in 1846, Challis and Galle discovered Neptune where it had been postulated to be. Predictions in this category may be based on empirical models that are not widely accepted by the majority of scientists. For example, astrological predictions are based on the objective motions of the planets, and in principle the results can be repeated by any qualified practitioner of the science of casting horoscopes. Saint Augustine has a convincing argument against the validity of this particular category of empirically based predictions that effectively places astrology in the subcategory of pseudoscience. In the example of horoscopes cast on the basis of the arrangement of the constellations for the Twins, Esau and Jacob, Saint Augustine points out that the astrologer's predictions in this case

> will not be true, because he would have consulted the same charts for both Esau and Jacob and would have made the same predictions for each of them, whereas it is a fact that the same things did not happen to them both. Therefore, either he would have been wrong in his predictions or, if his forecast was correct, he would not have predicted the same future for each. And yet he would have consulted the same chart in each case. This proves that if he had foretold the truth, it would have been by luck, not by skill.[95]

In the prophecies of both scientists and pseudoscientists, no divine intervention is required or necessary for the completion of the prediction or the making of it. Both are based on empirical observations. When a genuinely scientific prediction is fulfilled, it is consonant with natural law; when the prediction of a pseudoscientist – for example, an astrologer – is fulfilled, it is luck.

The second class of predictions is that in which the prophet receives a kind of intuitive, oracular, inner illumination that is wholly unrelated to any empirically derived model of reality. Under this category are predictions made on the basis of hunches, guesses, or imaginings. Shakespeare's soothsayer who bade Caesar to beware the Ides of March was probably in this category. As with the pseudoscientific predictions of the astrologer, whether such seemingly random utterances come to pass as predicted is more a matter of chance than of any special ability. Because of their affront to God in attempting to see beyond themselves and into the future, Dante lumps astrologers and soothsayers (or "diviners") together deep in the recesses of Hell, where, because of the nature of their sin, their heads are twisted completely around on their bodies. Unable to see in front of them, much less into the future, they walk through eternity facing backward.[96]

Neither the endogenous predictions of soothsayers nor the exogenous predictions of scientists and pseudoscientists require special divine dispensation whereby the predictor or prophet is empowered by God to foretell the future. Random predictive utterances based on hunches, guesses, imaginings, dreams, or visions, like scientific predictions foretelling an event consonant with an empirical model of nature and nature's laws, can never demonstrate a specially provident, divine intervention in the course of nature.

The third sense of "prediction" does signify specially provident, divine intervention. A prophet, in this sense, is "one of the sacred writers empowered by God to foretell future events."[97] In this sense, prophetic insight or ability transcends merely human predictions made by scientists and soothsayers. To be truly significant of God's special providence, a prophet's ability to see into the future must derive from God. A prophecy, in this sense, must be, as Locke noted, a miraculous event that transgresses the ordinary course of natural law. The prophetic predictions of Old Testament prophets may be broken down into two subcategories: (1) those that interpreters claim have already been fullfilled in history (historical prophecies); and (2) those that interpreters claim have not yet come to pass (prophecies of the future, such as the millennial prophecies of Daniel and Revelation).

It is the biblical interpreter or exegete who categorizes prophetic predictions as either "historical" or "future" prophecies – that is, as completed or not completed. In the first half of the eighteenth century, one of the most important subclasses of scientific predictions was that of biblical interpretation. Biblical interpretation may not appear to us today as akin in any way to the empirically derived, empirically testable predictions of physical science. However, William Whiston's main contribution to a comprehensive, fully articulated "Newtonianism" is his extension of the empirical method of Newton and the Royal Society into the realm of interpreting prophetic predictions uttered by divinely inspired prophets. A scientific exegete, he argued, can examine the genuine documents containing the prophetic predictions and their completion or lack of completion in the data of history. Feigned hypotheses have no place in the science of biblical interpretation, where, as in the physical sciences, the experimental method must reign supreme. Whiston claims that biblical interpretation is simply another application of the method used by the natural philosophers of the Royal Society, legal judges, and physicians. Whiston writes:

> Nor do I find that Mankind are usually influenc'd to change their opinion by any Thing so much, as by Matters of Fact and Experiment; either appealing to their own Senses now; or by the faithful Histories of such Facts and Experiments that appealed to the Senses of former Ages. And if once the Learned came to be as wise in Religious Matters, as they are

> now generally become in those that are Philosophical and Medical, and Judicial; if they will imitate the Royal Society, the College of Physicians, or the Judges in Courts of Justice . . . And if they will then proceed in their Enquiries about Reveal'd Religion, by real Evidence and Ancient Records, I verily believe . . . that the Variety of Opinions about those Matters now in the World, will gradually diminish; the Objections against the Bible will greatly wear off; and genuine Christianity, without either *Priestcraft* or *Laycraft,* will more and more take Place among Mankind.[98]

Whiston is somewhat equivocal about the degree of certainty that the biblical interpreter of completed "historical" prophecies can attain. Sometimes he indicates that complete certainty is beyond reach, and promises only probable conclusions: "And I faithfully promise... that upon the Appearance of such real Evidence, I will carefully consider it, and determine my Judgement on that Side of each Question, on which the *Momentum* shall appear to *Preponderate*."[99] But at other times Whiston indicates that it really does not matter, for the certainty of the Christian religion, if particularly obscure or problematic prophetic predictions cannot be certainly interpreted. In spite of "Errors in our Copies, or the Mistakes in our Histories, or the like Difficulties" that "have hitherto prevented our authentick Determination about them," nevertheless the momentum appears to preponderate, "to the undeniable Vindication and Confirmation of the certain Truth of the Old and New Testament."[100] Whiston's books on historically completed predictions, containing nearly four hundred prophetic predictions on one half of the page and the careful analysis of the historical record of their completion on the other half, are, in his eyes, casebooks of experiments that overwhelmingly verify the truth of revelation in an empirical fashion while ignoring hypotheses and metaphysics.

As for those prophetic predictions of the future that remain unfulfilled, the Newtonian exegete, as we saw in Chapter 3, does not have nearly the degree of certainty in interpreting their likely time and manner of completion as he does in interpreting the already completed "historical" prophetic predictions. Biblical interpreters are not themselves divinely inspired prophets miraculously "empowered" by God to foreknow the future. Still, the scientific biblical interpreter has justification, or, as Whiston states, "good reasons," to expect "the completion of those other prophecies whose periods are not yet come."[101] Newton, too, in his later years, as we have seen, was extremely cautious in himself predicting the actual times and means for the accomplishment of "future" prophetic predictions. But, like Whiston, he felt that he had good reason to hope that they would be completed:

> There is already so much of Prophecy fulfilled, that as many will take pains in this study, may see sufficient instance of God's providence; but then the signal revolutions predicted by all the holy Prophets, will at once both turn men's eyes upon considering the predictions, and plainly interpret them. Till then we must content ourselves with interpreting what hath already been fulfilled.[102]

Figure 5 is a diagram of the three types of prophetic predictions. An awareness of these distinctions and an understanding of the historical context of the status of prophetic predictions and the science of interpreting them make it easier to understand why Hume's claim that his argument against miracles also applies to prophecies was not an idle, uncomplicated remark when it was first published in 1748 in his *Philosophical Essays concerning Human Understanding* (retitled by Hume, in 1758, *Enquiry concerning Human Understanding*). The Newtonians were eager to promote the argument from prophecy because it seemingly evades the tension between a strong conception of generally provident natural order and law *and* the specially provident power of the creator to break that law. While emphasizing the empirical and repeatable nature of the "science" of biblical interpretation, most of the Newtonians (though not all) tended to ignore the fact that the essential nature of an Old Testament prophecy is necessarily miraculous and involves a break in or suspension of the laws of nature. So long as the necessarily miraculous nature of prophecy is ignored, the tension between naturalism and supernaturalism, between Newton's God and Newton's physics, is evaded.

But Hume permits no such ambiguity or evasion. Hume understands that for a completed "historical" prophecy to be demonstrative of special divine providence, it must be miraculous, which renders it automatically unbelievable. Completed "historical" prophecies offer no resolution to the tension between the traditional, specially provident, Christian God and the concept of the rigid, unbreakable laws of nature established by general providence. If one understands them properly, then completed prophecies are like other miracles that violate natural law as established in the course of human experience. As such, "historical" prophecies are simply too implausible to be believed.

The second section of the essay "Of Miracles" may be read as the completion of Hume's case against the Newtonian biblical interpreters. There Hume discredits the historical evidence that exegetes utilize to establish either the occurrence of miracles or, what amounts to the same thing, the completion (or "accomplishment") of prophetic predictions. Hume does not believe a word of Whiston's "faithful Histories of such Facts and Ex-

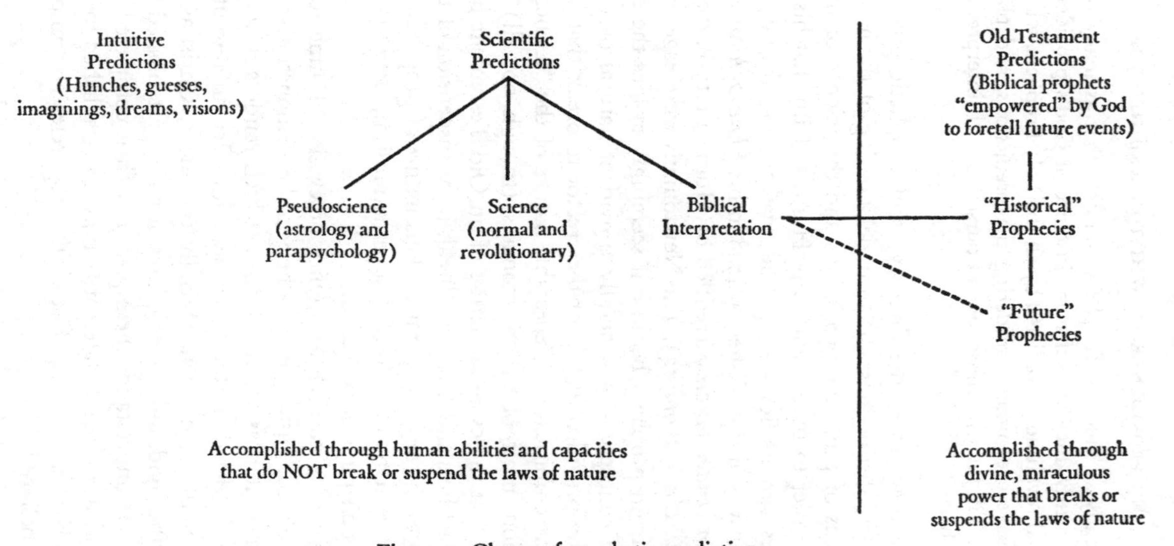

Figure 5. Classes of prophetic predictions.

periments that appealed to the Senses of former Ages," basically because they are usually contained in the same historical document (the Bible) as the prophecies themselves and are therefore untrustworthy, and also precisely because these "facts and Experiments" did appeal to the vulgar, unenlightened, non-Scottish ancients.

At a single stroke, Hume, as a philosophical historian, does away with the "science" of biblical interpretation.[103] As long as the Bible contains any miracle stories founded on human testimony and as long as it contains any stories of historically completed prophetic predictions beyond the capacity of scientific reason or mere chance to predict, it will be too implausible to be believed. Nothing can constitute adequate evidence for such violations of natural order and natural law. For similar reasons, in the *Dialogues concerning Natural Religion,* Hume dismisses "future" prophecies along with "historical" prophecies as simply beyond our experience and, hence, capacity:

> When we carry our speculations into the two eternities, before and after the present state of things; into the creation and formation of the universe; the existence and properties of spirits; the powers and operations of one universal spirit; omnipotent, omniscient, immutable, infinite, and incomprehensible: We must be far removed from the smallest tendency to scepticism not to be apprehensive that we have here got quite beyond the reach of our faculties.[104]

IV. Conclusion

Hume is the only writer in the eighteenth century to attack all the separate elements of the theory that scientist-theologians such as Whiston, Newton, and some members of the early Royal Society carefully crafted to demonstrate the existence of both general and special providence. Other wits and controversialists, such as Woolston,[105] attacked miracles. Collins, in a protracted dispute with Whiston, attacked prophecies. Wits such as Whitehead, Gay, Pope, Swift, and Walpole conflated everybody's positions and ridiculed them equally. But Hume combined serious attacks on belief in miracles and prophecies with an attack on the inference from the design argument of a generally provident deity. Hume's familiar religious skepticism in his essays takes on a new dimension when viewed against this historical context.

Hume's attack on prophecies not only combines with his attack on miracles to rule out specially provident divine intervention in world history; it also secularizes the future course of history. Hume removes more than a specially provident God when he points out that it exceeds the capacity of human beings to predict future history and that such miracles are in principle unbelievable. He also removes the anticipations of the future course of history in which the Newtonians of the "first" era of the Royal Society so ardently believed. Of Newton's attempt, in his *Observations upon the Prophecies of Daniel and the Apocalypse of St. John,* to interpret "historical" prophecies and to give us hope for the completion of those yet unfulfilled, Hume, the Newton of the mind, notes that we "never should pronounce the folly of an individual, from his admitting popular errors consecrated by the appearance of religion."[106]

Hume disconnects the frame of nature from a generally provident creator-designer and then eliminates any continuous, specially provident divine participation of the deity in the world around us. He thereby eliminates the prophetic past and the prophetic future.

When seen in the context of the apologetic concerns of the members of the "first" Royal Society, the sweeping nature of Hume's rejection of their synthesis of general and special providence is striking. In his coordinated attack on all the facets of their position, he far exceeds the casual ridicule of the scoffing dilettantes and wits, and by 1750 few members within the "second" era of the Royal Society coupled scientific research into physical phenomena with religious apologetics. When viewed in this context, we see Hume as a prophet of our own secular era. It is his prophetic vision of our modern secular world in which humanity is disconnected from God and God is disconnected from human history that constitutes Hume's uniqueness.

Even though Hume's immediate contemporaries, such as the Tory High Churchman Dr. Johnson, failed to perceive the devastating effect of Hume's critique of the Newtonian effort to retain a belief in special providence, more perceptive readers at a later date were not so oblivious. Echoing Hume, though not directly acknowledging him, the poet Shelley summarized the entire argument from prophecy and, with Hume, pointed out exactly where it fails:

> But prophecy requires proof in its character as a miracle; we have no right to suppose that a man fore knew future events from God, until it is demonstrated that he neither could know them by his own exertions, nor that the writings, pretending to divine inspiration should have really been divinely inspired; when we consider that the latter supposition makes God at once the creator of the human mind and ignorant of its primary powers, particularly as we have numberless instances of false religions, and forged prophecies of things long past, and no accredited case of God having conversed with men directly or indirectly.[107]

Increasingly, even the idea that a generally provident deity is a necessary inference from the theory of creation by divine fiat was challenged. By the end of the century, when Napoleon declared his displeasure with Laplace's effort to perfect Newtonian physics because it showed "that a chain of natural causes would account for the construction *and* preservation of the wonderful system" without any reference to God, Laplace made the famous reply, "God is a hypothesis of which I have no need."[108] Shelley encapsulates the modern point of view when he remarks, with admiring approbation, "The consistent Newtonian is necessarily an atheist"[109] – a comment that would have brought a vigorous (and no doubt lengthy) retort from Newton's controversial disciple, William Whiston.

NOTES

INTRODUCTION: ERKENNEN AND VERSTEHEN IN THE HISTORY OF PHILOSOPHY, AND WILLIAM WHISTON'S "NEWTONIANISM"

1 Albert William Levi, *Philosophy as Social Expression* (Chicago: University of Chicago Press, 1974), Chapter 1, especially pp. 8–16.
2 Ibid., p. 16.
3 Frank E. Manuel, *Isaac Newton, Historian* (Cambridge, Mass.: Harvard University Press, 1963), p. 143.
4 Richard S. Westfall, *Never at Rest: A Biography of Isaac Newton* (Cambridge: Cambridge University Press, 1980), p. 651.
5 Maureen Farrell, *William Whiston* (New York: Arno Press, 1981), p. 35. This work was originally presented as a doctoral dissertation at the University of Manchester in 1973 under the title "The Life and Work of William Whiston."
6 Robert H. Hurlbutt, *Hume, Newton, and the Design Argument* (Lincoln: University of Nebraska Press, 1965), pp. 37–8. Hurlbutt also sees that Whiston's attempt to use the new science for theological purposes is both shrewder and more historically significant than such splenetic nineteenth-century critics as J. M. Robertson and Leslie Stephen believed (ibid., pp. 38–9).
7 Hélène Metzger, *Attraction universelle et religion naturelle chez quelques commentateurs anglais de Newton,* deuxième partie (Paris: Hermann, 1938), pp. 95–103.
8 Manuel, *Newton, Historian,* p. 143.
9 Westfall, *Never at Rest,* pp. 647–53, 834–6.
10 Margaret C. Jacob, *The Newtonians and the English Revolution, 1689–1720* (Ithaca: Cornell University Press, 1976), pp. 130–4, 267–8.
11 Whiston, *Memoirs of the Life and Writings of Mr. William Whiston. Containing Memoirs of Several of his Friends also,* 2nd ed., 2 vols. (London, 1753), 1:259. Unless otherwise indicated, all citations from Whiston's *Memoirs* will be from this edition.
12 Whiston, *Astronomical Principles of Religion, Natural and Reveal'd* (London: 1717), p. 133.
13 See Berkeley to John Percival, September 6, 1710, in *The Correspondence of George Berkeley and Sir John Percival,* ed. Benjamin Rand (Cambridge, Mass.: Harvard University Press, 1914). It is at least possible that Berkeley learned of this approach to Genesis directly from Whiston; he took a copy of Whiston's *New Theory* with him to New England in 1728. See Chapter 1, note 94.
14 Whiston, *Memoirs,* p. 38.
15 Whiston, *A New Theory of the Earth, From its Original, to the Consummation of all Things, Where in the Creation of the World in Six Days, the Universal*

Deluge, And the General Conflagration, As laid down in the Holy Scriptures, Are Shewn to be perfectly agreeable to Reason and Philosophy. With a large Introductory Discourse concerning the Genuine Nature, Stile, and Extent of the Mosaick History of the Creation (London, 1696). These postulates, which summarize the entire "Introductory Discourse," are found on p. 95 of this separately paginated essay, which remained unchanged in all subsequent editions.

16 Whiston, *The Accomplishment of Scripture Prophecy* (London, 1708), p. 15.

17 Jacob, *Newtonians and the English Revolution*, p. 269.

18 Richard S. Westfall, "Isaac Newton's 'Theologiae Gentilis Origines Philosophicae,'" in *The Secular Mind: Transformations of Faith in Modern Europe: Essays Presented to Franklin L. Baumer, Randolph W. Townsend Professor of History, Yale University*, ed. W. Warren Wagar (New York: Holmes & Meier, 1982), pp. 30–1.

1. THE TEMPER AND TIMES OF A NEWTONIAN CONTROVERSIALIST

1 I wish to state at the very start that I consider the *Memoirs of the Life and Writings of Mr. William Whiston; Containing Memoirs of Several of his Friends also; Written by himself* and Whiston's other personal records of his life to be historically accurate. These *Memoirs*, composed between Whiston's seventy-ninth and eighty-second years, were first published in 1749 and were revised when Whiston was eighty-three or eighty-four, while he was preparing a second edition to include his 1750 London lectures. This second edition appeared posthumously in 1753. Unless otherwise noted, citations are from this edition. Of these *Memoirs*, Leslie Stephen observes that "the charm of his simple-minded honesty gives great interest to his autobiography; though a large part of it is occupied with rather tiresome accounts of his writings." (*Dictionary of National Biography* [*DNB*], s.v. "Whiston, William [1667–1752]"). I agree completely with this assessment of Whiston's autobiography, and because it so accurately reflects the temper of his times and Whiston's own temperament, I have made it the basis for the following sketch of his intellectual career. For another approach, see Maureen Farrell, *William Whiston* (New York: Arno Press, 1981), Chapter 1. The second, posthumous edition of the *Memoirs* was published by John Whiston (Whiston's son, a printer) and B. White in Fleet Street in 1753. It contains a preface, probably by Whiston's son, which notes that the work was being reprinted in its revised form because the 1749 edition had been sold out and the demand for the book still continued. The writer of the preface justly observed that "the strict integrity and sincerity of the Author, were probably the strongest motives for the favourable Reception which the publick have given to these Memoirs; and those will rather encrease, than be diminished by time" (n.p.). My view of the accuracy of Whiston's *Memoirs* is apparently shared by J. E. McGuire and Martin Tamny, who utilize Whiston's *Memoirs* to show that Newton accepted the existence of Cartesian vortexes as late as the 1680s. See their edition of *Certain Philosophical Questions: Newton's Trinity Notebook* (Cambridge: Cambridge University Press, 1983), p. 305.

2 Whiston, *Memoirs*, 1:5.

3 Ibid., p. 17. Whiston observes, as an example: "When I was become so vapoured and timourous at home, that I was ready to faint away if I did but go a few stones cast from our own house, my father observed it; and fearing the increase

of that distemper . . . he forced me to walk with our clerk, *John Flavell,* four miles on a frosty morning . . . Accordingly, when I found myself pretty well, both on my journey and return, I began to take a little more courage; and that degree of melancholy wore off, though a lesser degree of it always has, and I suppose always will, continue with me all the days of my life" (ibid., p. 18).

4 Ibid., p. 22. Whiston mentions hearing, in May 1688 or 1689, a brilliant sermon by a "clergyman, to me unknown; but, by his Hood seemed to be a Doctor of Divinity; I now suppose it was Dr. Prideaux!" In the circumstances – that is, the fear of "Popery and Persecution" – this divine created a stir with a carefully crafted sermon advising that "if Danger should approach, we should hold fast to our [Protestant] Bishops, as the most likely Way to escape the Dangers we might be in" (ibid.). Whiston was present with the strongly royalist senior fellow of his college, Dr. Nathaniel Vincent, whom he had accompanied into Norfolk "on account of my Health." Dr. Vincent's "Court-Sermon" was ill received by this company.

5 Ibid., p. 27.

6 Whiston went to ask Lloyd to ordain him, armed with two sets of documents: his "College Testimonials" or letters of recommendation, and a letter of recommendation from a Mr. Langley, minister of Tamworth. Lloyd consented to ordain Whiston, remarking to Whiston "that he knew what College Testimonials were: And that had it not been for Mr. *Langley's* Letter, I might have gone away *re infecta*" (ibid., p. 28).

7 Ibid., p. 30.

8 Ibid., p. 32. "Publick schools" refers to the public lectures that all students at the university were required to attend.

9 *Astronomical Principles of Religion, Natural and Reveal'd* (London, 1717), p. 257.

10 Whiston, *Memoirs,* 1:34. Whiston refers to the advantages of "the present great Light" enjoyed by the "Moderns" in stemming the recent "Torrent of Scepticism." See "Reason and Philosophy no Enemies to Faith," in his *Sermons and Essays upon Several Subjects* (London, 1709), pp. 198–200. Whiston once asked Newton why he himself did not draw the religious implications from his system as "Dr. Bentley soon did in his excellent Sermons at Mr. *Boyle's Lectures;* and as I soon did in my *New Theory;* and more largely afterward in my *Astronomical Principles of Religion;* and as that Great Mathematician Mr. Cotes did in his excellent *Preface* to the later Editions of Sir. I. N.'s *Principia:* I mean for the advantage of Natural Religion, and the Interposition of the Divine Power and Providence in the Constitution of the World; His answer was, that 'He saw those Consequences; but thought it better to let his Readers draw them first of themselves: Which Consequences however, He did in great measure draw himself long afterwards in the later Editions of his *Principia,* in that admirable *General Scholium* at its conclusion; and elsewhere in his *Opticks*" (Whiston, *A Collection of Authentick Records Belonging to the Old and New Testament,* 2 vols. [London, 1728], vol. 2, Appendix IX, "A Confutation of Sir Isaac Newton's Chronology," pp. 1073–4).

11 *Memoirs,* 1:38.

12 Ibid.

13 Ibid., p. 113. An account of Whiston's similar strenuous exertions to procure the Clare Fellowship for the sober Richard Laughton in 1697 is given in W. J.

Harrison, *Life in Clare Hall Cambridge, 1658–1713* (Cambridge: Heffer & Sons, 1958), pp. 86–8).

14 *Memoirs,* 1:109.

15 Ibid., p. 12. For an account of Whiston's close friendships with Dissenters, see A. G. Mathews, *Calamny Revised* (Oxford: Oxford University Press, 1934), pp. 523–4.

16 *Memoirs,* 1:110.

17 Ibid., p. 250.

18 Ibid., pp. 114–15.

19 Ibid., p. 118. Whiston published Newton's algebra lectures under the title *Sir Isaac Newton's Arithmetica Universalis.* Westfall notes Newton's petty criticisms of Whiston's edition: He objected to the title (which was identical to the title Newton had placed on his own abortive revision of the manuscript in 1684) and to the running heads at the top of each page, e.g., "Algebrae elementa." Although Newton claimed that he had to republish this work to correct Whiston's many errors, when he did so in 1722 the changes he actually made were reorderings of his own manuscript and not changes to Whiston's edition. See Westfall, *Never at Rest: A Biography of Isaac Newton* (Cambridge: Cambridge University Press, 1981), p. 649. Whiston acknowledges with easy grace that his edition of Newton's lectures was superseded by that of "Mr. Machin, one of the secretaries of the royal society, by the author's later desire or permission" (Whiston, *Memoirs,* 1:118).

20 Whiston, *Memoirs,* 1:117. These lectures on astronomy were translated into English and printed in 1710; a second edition appeared in 1728. Whiston's lectures on Newtonian mathematical philosophy were published in 1710 as *Praelectiones Physico-Mathematicae;* they were translated and printed in 1716.

21 Whiston, *Memoirs,* 1:116.

22 Ibid., p. 118. Whiston notes that half of the lectures were by him and half by Cotes. He goes on to observe that "I esteem mine so far inferior to his, and many later books and courses relating to such matters being become common, I cannot prevail with myself so much to revise and improve them, as they ought to be before they are fit for publication" (ibid., pp. 118–19). Whiston adds that "the present Duke of *Argyle* took a copy of them long ago, when he had gone through our course" (ibid., p. 119).

23 From William Stukeley, *The Family Memoirs of the Rev. William Stukeley, M.D.;* cited in A. E. Clark-Kennedy, *Stephen Hales, D.D., F.R.S. An Eighteenth Century Biography* (Cambridge: Cambridge University Press, 1929), p. 19. Hales and Stukeley also attended Cotes's astronomical lectures.

24 Whiston, *Historical Memoirs of the Life of Dr. Samuel Clarke. Being A Supplement to Dr. Syke's and Bishop Hoadley's Accounts. Including Certain Memoirs of several of Dr. Clarke's Friends* (London, 1730), pp. 11–12.

25 Ibid., p. 13. Whiston is clear that such doubts about the Athanasian creed were quite new to him, "as a Sermon of mine Preach'd upon Christmas Day about 1704, at great St. *Bartholomews,* if now Extant, would Witness." Whiston goes on to add, "Whether Mr. *Newton* had given Mr. *Clarke* yet any intimations of that nature; for he knew it long before this time; or whether it arose from some enquiries of his own I do not directly know: tho' I incline to the latter" (ibid.). Just when Whiston learned that Newton shared his and Clarkes's Arian views is not known, but to judge from this text it was probably around 1705.

Writing in 1728, Whiston states that Newton's Arianism "was occasionally known to those few who were intimate with him all along; from whom, notwithstanding his prodigiously fearful, cautious, and suspicious temper, he could not always conceal so important a Discovery" (*Authentick Records,* 2:1077).

26 Whiston, *An Historical Preface to Primitive Christianity reviv'd. With an Appendix Containing An Account of the Author's Prosecution at, and Banishment from the University of Cambridge* (London, 1711), p. 2. With a preliminary draft of the evidence from "primitive" Christianity in hand, Whiston went to London to show it to a Dr. Bradford, Benjamin Hoadley, Samuel Clarke, and a Mr. Sydal, who, he says, already "shrewdly" suspected that the common, orthodox notion was false (ibid., p. 6; *Memoirs of Clarke,* p. 15).

27 Whiston, *Historical Preface,* p. 7.

28 Ibid., pp. 7, 13.

29 Whiston sent the same letter, dated July 17, 1708, to both archbishops. He included a copy of his letter and copies of their replies – from which this text and the two following are taken – in his *Historical Preface,* pp. 15–18. Whiston sent the letter to Archbishop Sharp via Samuel Clarke. In a separate cover letter to Clarke he asked him to peruse the letter to Sharp and advised him to renounce his caution: "Above all, act openly; advise with Sir *Isaac Newton:* and, if you can do it with a safe conscience in that sense, declare at the time of subscription, that you sign them as Articles of Peace, which you will never oppose by Preaching or Writing, and no farther" (*Memoirs of Clarke,* p. 17).

30 Whiston, *Historical Preface,* p. 18. The archbishop did say that he could not judge Whiston's research until he had perused it and that he would rather see it "in Writing than in Print."

31 Ibid., p. 19.

32 Ibid., p. 21. There is a difference between Lloyd's manuscript copy of his letter and Whiston's printed version of Lloyd's letter. Whiston's printed version ends, "God knows I desire nothing else but your good, and to keep you from doing Hurt to the Church. I beseech God that these Notices I have given you may have the effect I design by them." The manuscript letter of Lloyd's copy ends, "But knowing so much as I do of the things before mentioned, I fear you. And therefore I hold it my duty to do a friend's part in warning you whither you are going" (British Library manuscripts: additional manuscript 24197, ff. 16–47).

33 Whiston, *Historical Preface,* p. 7.

34 Whiston, *Sermons and Essays.* In the *Memoirs,* p. 127, Whiston writes that since the publication of his "learned friend" Mr. Jackson's annotated edition of the essay "De Trinitate" in 1728, he desires it to be omitted from all future editions. Hence the note in the British Museum edition of the *Sermons and Essays* in Whiston's hand that reads "Omit this Treatise in Future Editions." He is not recanting his antitrinitarianism but deferring to a more useful edition. Also contained in these *Sermons and Essays* is an essay recommending the establishment of charity schools; one on the restoration of the Jews; and one, "Reason and Philosophy no Enemies to Faith," in which Whiston argues against the eternity of hell torments. This latter essay was later expanded into a larger pamphlet that increased Whiston's reputation for heterodoxy. See D. P. Walker, *The Decline of Hell* (London: Routledge, & Kegan Paul, 1974), Chapter 6.

35 Whiston, *Memoirs,* 1:131. Whiston's Primrosian attitude is made even plainer in this text: "When I had drawn up this Advice for the Study of Divinity,

(which has been since in part made publick;) and began to speak of it to some Friends, and freely to declare my thoughts about the Doctrine of the Trinity, I was immediately made sensible what a nice Point I was engag'd in; and what a noise, and bustle, and *odium,* and perhaps Persecution, I should raise against my self, if I ventur'd to talk and print at that rate; and how I and my Family would probably be ruin'd by such a Procedure. As to my own worldly Interest, and that of my Family, I very well knew the Duty of a Christian; and all along firmly resolv'd that such Arguments should have no Influence upon me, nor in the least discourage me from speaking and Writing the Truths of *Christ Jesus,* when upon a thorough Examination I found them to be such" (*Historical Preface,* p. 3).

36 Whiston, *Memoirs,* 1:131–2.

37 Whiston, *An Account of the Author's Prosecution at, and Banishment from the University of Cambridge* (London, 1711), p. 2. This work was first printed as an appendix to the *Historical Preface* and was reprinted in 1718 upon the occasion of Bentley's prosecution and suspension from the university.

38 Ibid., p. 10. This position held and taught by Whiston was contrary to the first, second, and fifth of the Thirty-nine Articles as well as to the Nicene and Athanasian creeds. The evidence cited in this charge is taken from the English sections of the *Sermons and Essays,* e.g., pp. 213–15, pp. 276–8. Another of the charges was based on Whiston's one and only erratum note to the *Sermons and Essays,* which changed the wording of the doxology from "Three Persons and One God" to simply "in the Holy Ghost" (ibid., p. 5; cf. *Sermons and Essays,* pp. 123, 412; and *Memoirs,* 1:116).

39 Whiston, *Memoirs,* 1:132.

40 Ibid., p. 145.

41 Ibid., pp. 140–1 (brackets are in the original). King was one of Whiston's defense counsels in the trial before the Court of Delegates. When Whiston learned that Dr. Harris, chaplain to the Prince of Wales in the reign of George I, would not inform the prince that a maid of honor to the Princess of Wales was a member of the Hell Fire Club, because he expected preferment and did not want to get involved, Whiston took on the job himself. He visited a friend of the princess, Lady Gemmingen, to whose brother Whiston was teaching mathematics. He informed her about the scandal, and Lady Gemmingen informed the princess, "but upon enquiry, nobody would confess themselves guilty: tho' the thing at that time was but too notorious. Only some stop was, I suppose, put to that infamous club for that time. But O, what a sad, but prevalent topick am I now come to! *The Expectation of Preferment: More Preferment!* the grand thing commonly aimed at both by clergy and laity; and generally the utter ruin of virtue and religion among them both!" (ibid., p. 135).

42 Some representative titles from this extensive pamphlet warfare are R. Smalbroke, *The new Arian reprov'd: or, a vindication of some reflections on the conduct of Mr. W., in his revival of the Arian heresy: from Mr. W.'s late animadversions, inserted in his Second Reply to Dr. Allix . . .* (London, 1711), and Whiston, *A second Reply to Doctor Allix, with two postscripts: the first to Mr. Chishull; the second to the author of the Reflections on Mr. Whiston's conduct* (London, 1711). A nonacademic polemic attacking Whiston's avowed Arianism and public lecturing was printed in London in 1714; it departed radically in tone from the more sober and scholarly attacks of Whiston's orthodox opponents.

This work was entitled *Will-with-a-Wisp; Or, the Grand Ignis Fatuus of London. Being a Lay-Man's Letter to a Country Gentleman, concerning the Articles lately exhibited against Mr. Whiston.* The anonymous author complains that London's youth follow Whiston about, listening to him speak like a latter-day Socrates, and thus "contract ill habits and Wrong Notions in Religion, as well as neglect their other worldly *Employments,* by following such admir'd designing, Arian Lectures in vogue" (p. 57).

43 Whiston, *Memoirs,* 202–3; *Memoirs of Clarke,* pp. 86–96.

44 *Memoirs,* 1:233.

45 W. H. Wilkins, *Caroline the Illustrious. Queen-Consort of George II. and sometime Queen Regent. A Study of her Life and Time* (London: Longmans, Green, 1904), p. 246. Whiston only hints at this possible connection with Caroline, as the following tantalizing anecdote reveals. Whiston was told in 1736 that "the late duke of *Somerset,* a great *Athanasian,* once forbad his chaplain to read the *Athanasian* creed, (which I imagined was occasioned by a suggestion from the queen; to whom I had complained, that altho' she was queen, that creed was not yet laid aside)" (*Memoirs,* 1:299). Whiston also mentions stopping in and trying to see Addison on his deathbed, while "passing to the queen at Richmond" (ibid., p. 258).

46 Wilkins, *Caroline,* p. 246. This story is repeated in the *Dictionary of National Biography* article on Whiston.

47 Whiston, *Memoirs,* 1:290–1.

48 Ibid., p. 151.

49 Newman to Steele, August 10, 1713. Rae Blanchard has described the discovery of an extensive correspondence between Newman and Steele, which includes this letter, in the manuscript Letter Books of the London SPCK. See Rae Blanchard, "Richard Steele and the Secretary of the SPCK," in *Restoration and Eighteenth-Century Literature,* ed. Carroll Camden (Chicago: University of Chicago Press, 1963), pp. 287–95.

50 Addison's father, while dean of Litchfield, had laid hands upon Whiston as presbyter at his ordination. Whiston recounts that the younger Addison was raised for the ministry but diverted into a political career by Lord Halifax and Lord Chancellor Somers because "his parts appeared so promising." But Addison's early theological training, Whiston says, was nevertheless reflected in the "*Saturday's* papers, in his famous *Spectator,*" which were generally on religious subjects and "were intended originally for sermons." Whiston claims that in his latter years Addison read the ancient fathers of the first three centuries and composed his *Discourse of the Christian Religion.* Whiston also claims to have collected all of Addison's references in this work to "several ancient testimonies" that were not found among Addison's own papers after his death and says that he had endeavored to get them published in future editions. Whiston delivered this copy of Addison's "ancient testimonies" to Addison's bookseller, Mr. Tonson, "tho' neither he nor [Addison's] family have yet published it, in these twenty-seven years time, to the disappointment of myself, and all inquisitive readers . . . nor have I kept any copy of them myself. So that if Mr. *Tonson* suppresses that paper, or has lost it, it is intirely lost both to me and to the publick" (*Memoirs,* 1:256).

51 Whiston, *Memoirs,* pp. 257–8. Steele gave notice in the *Englishman,* no. 42, (January 26, 1714), soon after the start of the lecture series, that because of their

popularity the lectures were being moved from Button's "to a larger Room close by at Mr. Dale's, an upholsterer over the Corner of the nearest Piazza in Covent Garden." See *"The Englishman": A Political Journal by Richard Steele,* ed. Rae Blanchard (Oxford: Clarendon Press, 1955), p. 436. Earlier Steele had printed an advertisement announcing the start of "a course of Philosophical Lectures on Mechanics, Hydrostatics, Pneumatics, Opticks. . . . This course of experiments is to be performed by Mr. William Whiston and Francis Hauksbee, the nephew of the late Mr. Hauksbee" (*Englishman,* no. 42, January 9, 1714, in Blanchard, *The Englishman,* p. 440). These advertisements apparently refer to the same series of lectures. On the evidence, they are the first public lectures Whiston delivered in London.

52 See F. W. Gibbs, "Itinerant Lecturers in Natural Philosophy," *Ambix* 2 (June 1960):111–17.

53 Whiston, *Memoirs,* p. 258. Stanhope discussed with Whiston the fulfillment of biblical prophecies in terms of such present-day political events as the defeat of the Turks at Corfu. See Whiston, *Literal Accomplishment of Scripture Prophecies* (London, 1724), p. 97.

54 Whiston, *Memoirs,* 1:204–5. Whiston invented an instrument that he called a "Copernicus" to aid in the calculation of the dates of future eclipses (ibid., p. 157).

55 Ibid., pp. 283, 1:332–3; the lectures were apocalyptic, because for Whiston "Ezekiel's Temple" refers to "the Jews *future* Temple." See *Authentick Records,* p. 1070; *Memoirs of Clarke,* p. 149. Marjorie Nicolson and G. S. Rousseau note in their *"This Long Disease, My Life": Alexander Pope and the Sciences* (Princeton: Princeton University Press, 1968), pp. 113–14, a problem in the interpretation of a one-page set of verses dating from 1727 in the Library of the Royal Society of Medicine and inscribed, "by John Arbuthnot." The last stanza reads, "WHISTON, much plainer than his Creed, / Those Beasts in Scripture saw; / But as the Story proves indeed / It was APOCHRYPHA." In the second volume of his *Memoirs* (second edition) Whiston had related the story of the "Rabbit Woman" Mary Toft as the fulfillment of the prophecy of 2 Esdras 5:9 concerning monstrous births. In April 1726, when she was five weeks pregnant, Mary Toft was surprised while working in a field near her Surrey home by a rabbit. In September she miscarried and claimed to have given birth to at least seventeen rabbits over the next several days. This story was investigated by two surgeons to George I, Nathanael St. André and Cyriacus Ahlers, and they declared it to be true. Eventually Mary confessed it had been a fraud and was committed to the Bridewell Prison in Tothill Fields, London. Whiston accepted her original story and rejected her recantation and, in the 1753 *Memoirs,* wove it into his tapestry of fulfilled biblical prophecies (*Memoirs,* 2:108–22. See Figure 4.) Nicolson and Rousseau pose the question of how Whiston's views, which were not published until 1753, could have been known to the witty "Arbuthnot," who pegged these views so accurately in 1727. The answer lies in Whiston's public lectures of 1726 about the events preceding the rebuilding of the Temple. This is the context of Whiston's theory of the Mary Toft case, which was published in 1753; and I have no doubt that it was part of his public lectures on the Temple in 1726. Arbuthnot satirized Whiston's apocalyptic lectures around 1729 in his "True and Faithful Narrative of What pass'd in London during the general Consternation of all Ranks and Degrees of Mankind," published in

Swift's and Pope's *Miscellanies. The Third Volume* (London, 1732), pp. 255–76 (see Chapter 5).

56 Whiston, *Memoirs,* 1:17. In a letter of April 21, 1736, to his son George, Whiston sympathizes with the "deep melancholy" that his son's "intense studies" have brought about in him. He adds, "I well remember that when I have been long and greatly melancholy in former years, let me only get on horseback, and I found myself better immediately." Cited in Farrell, *Whiston,* p. 38.

57 Whiston, around 1724, was mathematics tutor "to a very ingenious *German* youth, the baron Gemmingen" (*Memoirs,* 1:270; see note 41).

58 Addison, *The Guardian,* 2 vols. (London, 1714), 2:91. Steele also aided Whiston, as usual, by printing a letter of Whiston's defending his and Ditton's method for determining longitude as "certain, easie and practicable." See the *Englishman,* no. 29 Dec. 7, 1713). Earlier, writing in the *Spectator,* no. 428 (July 11, 1712), Steele had invited contributors "from the late noble Inventor of the Longitude, to the humble Author of Straps for Razors" to "take their Turns in my Paper." This reference may or may not refer to Whiston.

59 A report to the House committee found in Whiston's papers in the Barker manuscripts at the Leicestershire Record Office records that Cotes thought that "the project was Right in the Theory near the Shore, and the practicable part ought to be Experimented." In the same document, Halley says that this method "did seem to him to consist of many particulars which first ought to be experimented before he could give his Opinion." These texts are cited by Farrell, *Whiston,* Chapter 3, which gives an admirable account of Whiston and the longitude problem.

60 See Derek Howse, *Greenwich Time and the Discovery of the Longitude* (Oxford: Oxford University Press, 1980), pp. 46–56, and Westfall, *Newton,* pp. 834–5. Westfall summarizes Whiston's own account of the meeting that was published in Whiston's historical preface (dated separately, 1742) to his *Longitude Discovered by the Eclipses, Occultations and Conjunctions of Jupiter's Planets* (London, 1738), pp. ii–vii.

61 William Whiston and Humphrey Ditton, *A New Method for Discovering the Longitude both at Sea and on Land* (London, 1714).

62 Whiston's dipping needle method featured six- to eight-foot iron needles and maps with lines of varying degrees of magnetic attraction as determined by surveys. By comparing the angle of the needle with the chart, one's position might be established. The first such isogonic chart had been produced by Edmond Halley in 1700 subsequent to his Atlantic voyages as captain of H.M.S. *Paramore.* A group of "publick spirited" men, Whiston's "particular friends," raised by subscription over four hundred seventy pounds for Whiston's support while he was working on this method. King George I gave one hundred; the Prince and Princess of Wales gave eighty-eight; Sir Joseph Jekyll, thirty-one; the duke of Montague, twenty-one; and Martin Folkes, secretary of the Royal Society and the man who proposed Whiston for membership in 1716, over five pounds.

63 Eric Gray Forbes, "The Origin and Development of the Marine Chronometer," *Annals of Science* 22, no. 1 (March 1966), and Howse, *Greenwich Time,* pp. 67–72.

64 Westfall suggests that both Newton's temperament and the apparent theft by Whiston's nephew caused the rift. See *Newton,* pp. 652–3. Frank E. Manuel be-

lieves the cause was a combination of Whiston's enthusiasm for martyrdom and Newton's desire to avoid public debate. See Manuel, *Newton, Historian* (Cambridge, Mass.: Harvard University Press, 1963), pp. 143, 175, and his *Portrait of Isaac Newton* (Cambridge, Mass.: Harvard University Press, 1968), Chapter 13, "Newton as Autocrat of Science."

65 Whiston, *Memoirs*, 1:249–51. Whiston also notes their clash of temperaments in these terms in 1728: "What cautious Temper and Conduct on his, or what openness of Temper and Conduct on my Side, or what other Accidents occasion'd any Interruptions in that Favour and Friendship, 'tis not perhaps proper for me to say" (*Authentick Records*, p. 1071). Obviously, twenty years later, while writing his memoirs, Whiston thought it proper to expand his earlier account. Whiston was nominated by Martin Folkes for membership in the Royal Society in 1716, but the council apparently ignored the nomination completely. See *Journal Book Copy of the Royal Society* 11, p. 124 (cited in Westfall, *Newton*, p. 653). There is a problem in Whiston's account. Royal Society records show that he was proposed only once – by Folkes in 1716. Yet Whiston says that "on or about . . . 1720, I take it to have been," Sloane proposed him. Either he was proposed a second time by Sloane in 1720 and we have no record of it, or, writing in 1748, Whiston mistakes both the year (his phrasing is somewhat tentative) and the name of the person who proposed him. The latter hypothesis is most likely.

66 Whiston, *Memoirs*, 1:35–6.

67 Ibid., pp. 250–1.

68 Ibid., p. 284.

69 See Locke to Molyneaux, February 22, 1696, in *Some Familiar Letters Between Mr. Locke and Several of his Friends* (London, 1708), pp. 176–7. Whiston includes this letter in his *Memoirs*, 1:35; see also "The best Method of Studying, and Interpreting the Scriptures," in *The Remains of John Locke Esq . . . Publish'd from his Original Manuscripts* (London, 1714), p. 14.

70 Cited by Nicolson and Rousseau, *This Long Disease*, p. 138; they argue for a date of 1708 instead of 1707.

71 Cox's letter is dated December 23, 1714. See C. J. Rawson, "Parnell on Whiston," *Bibliographical Society of America Papers* 57 (1963): 91–2. Rawson found this letter in the British Museum collection of manuscripts. Cox goes on to copy out a shortened version of Parnell's verses that appeared in 1732 as part of Swift's, Pope's, Gay's, and Arbuthnot's *Miscellanies. The Last Volume* (London, 1727), pp. 172–3. Entitled "Ode, for Musick, on the Longitude," the first stanzas read: "The Longitude mist on / By wicked *Will. Whiston* / An not better hit on / By good Master *Ditton*" and "So *Ditton* and *Whiston* / May both be bep-st on; / And *Whiston* and *Ditton* / May both be besh-t on." Variations of this pattern are repeated in the next two stanzas. It was following the public circulation of this unsavory ad hominem attack that Steele published Whiston's and Ditton's letter of defense in the *Englishman*, no. 29 (Dec. 10, 1713), where the authors noted that it was quite probable that the editor might have "heard that Proposal of ours . . . ridicul'd" (Blanchard, "*The Englishman*," p. 119).

72 Abraham de la Pryme, *The Diary of Abraham de la Pryme* (Ripon, 1869), p. 159. Cited in Roy Porter, *The Making of Geology* (Cambridge: Cambridge University Press, 1977), p. 87.

73 Jonathan Swift, *Gulliver's Travels*, ed. Louis A. Landa (London: Methuen, 1960)

Part III, Chapter 10, p. 170. Gulliver speculates that if he were to become an immortal, like the "Struldbruggs," he "should then see the discovery of the longitude, the perpetual motion, the universal medicine, and many other great inventions brought to the utmost perfection." In a letter to Archbishop King in 1712, Swift complains that "a projector has lately applied to me to recommend him to the Ministry about an invention for finding out the longitude. He has given in a petition to the Queen by Mr. Secretary St. John. I understand nothing of the mathematics; but I am told it is a thing as improbable as the philosopher's stone, or perpetual motion" (*The Correspondence of Jonathan Swift, D.D.*, 6 vols., ed. F. Elrington Ball [London: G. Bell, 1910–14], 1:324). The identity of this particular "projector" is unknown.

74 John Arbuthnot, *An Examination of Dr. Woodward's Account of the Deluge* (London, 1697), Preface, and Arbuthnot to Swift, July 17, 1714, *Correspondence of Swift*, 2:186.

75 Swift to Arbuthnot, July 25, 1714, *Correspondence of Swift*, 2:196.

76 *Gulliver's Travels*, Part III, Chapter 2, p. 132.

77 Pope to Caryll, August 14, 1713, in *The Correspondence of Alexander Pope*, 5 vols, ed. George Sherburn (Oxford: Oxford University Press, 1956), 1:185; cited in Nicolson and Rousseau, *This Long Disease*, p. 137. This letter was apparently copied by Pope and sent to Addison on December 14, 1713 (*Correspondence of Pope*, pp. 201–3). Nicolson and Rousseau are eager to document Pope's interest in science and construct a case that Pope was profoundly influenced by the science of his day, that he was transported into the "vast abyss of eternity" by attendance at a series of lectures that they believe Whiston may have given in the summer of 1713. The evidence from Whiston's *Memoirs* and from Newman's letter to Steele cited in Blanchard, "Richard Steele and the Secretary of the SPCK" (see note 49), provides no textual evidence for supposing that Whiston lectured prior to January 1714. These lectures are the only lectures advertised by Steele in the *Englishman* in apparent fulfillment of his pledge to Newman to help the banished professor. Whether Pope attended these January 1714 lectures is unknown. What Pope says in his letter to Caryll of August 1713, and apparently four months later in his letter to Addison of December 1713, is that Whiston affected him profoundly in "dialogues," not in public lectures. The influence remains pronounced in any case. If the letter to Caryll, which was only printed by Pope in the 1730s, is a copy of the letter to Addison – that is, if the date is wrong – it may be that Pope did attend Whiston's first lectures. Recently Michael Hoskin has speculated about how Whiston's astronomical writings may have influenced the noted eighteenth-century astronomer Thomas Wright of Durham. See Hoskin, *Stellar Astronomy: Historical Studies* (Chalfont St. Giles: Science History Publications, 1982), pp. 101–2.

78 Whiston, *Memoirs*, 1:127.

79 *Ibid.*, p. 178.

80 T. L. Bushell, *The Sage of Salisbury: Thomas Chubb, 1679–1747* (New York: Philosophical Library, 1967), pp. 8–10. Whiston later denounced Chubb for becoming a religious skeptic. Bushell puts this down to Whiston's jealousy over the larger salary Jekyll paid to Chubb (pp. 12–13). Whiston was undeniably always looking for ways to increase his income, but he also always viewed religious skeptics as dangerous opponents to be opposed at whatever cost.

81 Whiston, *Memoirs,* 1:305–9. Here Whiston publishes a letter he wrote to John Depee, dated April 20, 1738, in which he answered Depee's queries about Arianism, the forgeries of Athanasius, and baptism.

82 Another instance of Whiston's reputation and influence as an astronomical prognosticator is provided by James Logan, Penn's agent in America, a classicist and man of science, and one-time mayor of Philadelphia and governor of Pennsylvania. Logan was in London on proprietary business on May 11, 1724. In company with some members of Penn's family, Logan spent the day at Windsor in order to observe the solar eclipse. He chose that site because Edmond Halley, the Astronomer Royal, had predicted that the eclipse would be fully observable there. Halley was incorrect in this prediction. Whiston's competing chart of the transit of the earth's shadow was more accurate, and on his copy of it Logan noted that Whiston's prediction of the band of totality was "by much the truest." Whiston's account of his method and conclusions is contained in *The Calculation of Solar Eclipses* (London, 1724). This anecdote about Logan comes from *The Annual Report of the Library Company of Philadelphia for the Year 1982, Presented at the Annual Meeting, May, 1983,* (Philadelphia: Library Company of Philadelphia, 1983), p. 10. I am much indebted to the intrepid John Bidwell, the Reference/Acquisitions Librarian at the William Andrews Clark Memorial Library, who sent a copy of the Library Company's *Annual Report* to me.

83 Mrs. Thrale, *Thraliana: The Diary of Mrs. Hester Lynch Thrale (Later Mrs. Piozzi), (1776–1809),* 2 vols., ed. Katherine C. Balderston (Oxford: Clarendon Press, 1942), 1:383–4.

84 See Eric Rothstein and Howard D. Weinbrot, "*The Vicar of Wakefield,* Mr. Wilmont, and the 'Whistonean Controversy,'" *Philological Quarterly* 55, no. 2 (Spring 1976):225–40, for an interpretation of Whiston as the model for Goldsmith's vicar.

85 See Whiston, *Memoirs,* 1:244–5, and *Bibliothèque angloise* 3 (1718):410–41. This journal also kept close track of Whiston's theological tracts and positions: e.g., his Arian disputation with the earl of Nottingham (9 [1721]:64–93, 269–71); his *Essay Toward Establishing the True Text of the Old Testament* (10 [1721]: 290–333); his *Supplement* to this *Essay* (11 [1724]:74); his *Literal Accomplishment of Scripture Prophecies* (12:[1724]:62–87); and the critical responses to his "system" (11[1724]:87–120; 12 [1724]:463).

86 See Manuel, *Newton, Historian,* pp. 173ff.

87 George-Louis Leclerc, Comte de Buffon, "Du Système de M. Whiston," in *Histoire et théorie de la terre,* vol. 1, *Histoire naturelle* (Paris, 1749), Premier discours, Article II.

88 In 1759, Pierre Louis Moureau de Maupertius, mathematician, astronomer, and president of the Royal Academy of Sciences in Berlin, published his *Essai de cosmologie* in which he describes the expected effects of a comet's close approach to a planet. Around 1744, Joseph Jérôme Le Français de Lalande wrote on the same topic in *Reflexions sur les comètes qui peuvent approcher de la terre.* According to Jean Baptiste Joseph Delambre, this paper was widely believed to predict a collision between earth and a comet for the year 1775 and caused great panic in Paris. The police sought to calm these fears by publishing Lalande's complete text in 1774, but the crowds believed that the police and the author had conspired to omit the crucial catastrophic prediction in order to maintain calm

up to the moment of fatal impact. See Delambre, *Histoire de l'astronomie au dix-huitième siècle* (Paris, 1827), pp. 547–621. Whiston's original theory is summarized on pp. 51–60.

89 One of the first German writers to popularize Whiston's *New Theory* in Germany was Dethlef Cluver. Cf. Dethlevus Cluverus, *Geologia, sive Philosophemata de genesi ac structura globi terreni. Oder: Natürliche Wissenschaft, von Ershaffung und Bereitung der Erd-Kugel* (Hamburg, 1700), passim. In 1710, accord-to Whiston, "*Menkenius* a learned man in *Germany,* wrote to Dr. *Hudson,* the learned keeper of the *Bodleian* library at Oxford, to procure him an account of me; whose writings then made, as he said, a great noise in *Germany*" (*Memoirs,* 1:130). This text seems to go on to indicate an interest in Whiston's religious polemics as well, for Menkenius was specifically curious about Whiston's character.

90 "W. W.," *Nova Telluris Theoria, Neue Betrachtung der Erde, nach ihren Ursprung. und Fortgang biss zur Hervor bringung aller Dinge . . . Aus dem Englischen ubersetzt von M.M.S.V.D.M.* (Frankfurt, 1714).

91 Joannes Buddeus, *Historia ecclesiastica Veteris Testamenti ab orbe condito usque ad Christum natum, variis observationibus illustrata* (Halae Magdeburgicae, 1719), 2 vols., 1:62. (The first edition appeared in 1715.)

92 Andrew Dickson White, *A History of the Warfare of Science with Theology in Christendom,* 2 vols. (New York, 1910), 1:206. Cited in the invaluable Katharine Brownell Collier, *Cosmogonies of Our Fathers* (New York: Columbia University Press, 1934), pp. 112–13. The work referred to in this passage by John Heyn, rector of a university in Brandenburg, is *Versuch Einer Betrachtung über Die Cometen, die Sundflut und das Vorspiel des jungsten gerichts, Nach astronomischen Grunden und der heligen Schrift angestellet* (Berlin, 1742). Heyn also added prefaces to two similar works by his students, Balthasar Friderick Kunstmann and John Gotthilf Werder, the whole appearing as *Specimen Cometologiae Sacrae* (Leipsig, 1742).

93 Whiston, *Memoirs,* pp. 15–16. Steven S. Schwarzschild has tentatively identified one of these scholars as Raphaël Levi. See "Raphael Levi de Hanovre et la Frühaufklarüng juive," *Le dix-huitième siècle,* April 1981, p. 32.

94 Andrew Keogh, "Bishop Berkeley's Gift of Books in 1733," *Yale University Library Gazette* 8 (July 1933):14, 21. Berkeley had been informed of Whiston's difficulties at Cambridge by Sir John Percival. See Eamon Duffy, "'Whiston's Affair': The Trials of a Primitive Christian, 1709–1714," *Journal of Ecclesiastical History* 27, no. 2 (April 1976), p. 137.

95 See Perry Miller, "The End of the World," *William and Mary Quarterly* 8, Third Series (April 1951):182.

96 Ezra Stiles, 1727–95, studied science and theology at Yale and became the president of that college. In 1755 he took the post of chief Congregationalist minister at Newport because of its excellent private library and cosmopolitan religious composition. His unpublished papers reveal that Whiston exerted much influence on Stiles's thought. In Stiles's notebooks of 1744–5 from Yale College, he has copied from Whiston's *Astronomical Lectures, Astronomical Principles,* and *New Theory,* long sections pertaining to calculating velocities and celestial distances. From the "Introductory Discourse" to the latter work, Stiles also abstracts Whiston's arguments that the Mosiac history of creation refers solely to the earth. Stiles was interested in both what we should today refer to as Whis-

ton's science and his theology, including his Arianism and millennialism. See College Notebook, April 20, 1745, in *The Microfilm Edition of the Ezra Stiles Papers at Yale University* (New Haven: Yale University Library, 1978), reel 13, item 45, pp. 30, 40–1, 47, 48–9, 50–1, 52–3; and Bound Book Miscellaneous Volume, ibid., reel 14, item 147, pp. 252–9.

97 Melville, *The Writings of Herman Melville,* Harrison Hayford (Evanston, Ill.: Northwestern University Press and the Newberry Library, 1968–82), vol. 3, *Mardi and a Voyage Thither,* ed. Harrison Hayford, Hershel Parker, G. Thomas Tanselle (1970), p. 5. Melville illustrates the polar to subtropical course of a whaling ship by comparing it to "the Whistonian theory concerning the damned and comets." Cf. Whiston, *Astronomical Principles,* pp. 155–6.

98 DNB, s.v. "Whiston, William." See also Leslie Stephen, *English Thought in the Eighteenth Century,* 2 vols. (New York: Harcourt, Brace & World, 1962), 1:179. (This work was first published in 1876). Stephen writes: "With a childlike simplicity worthy of the Vicar of Wakefield, he was ready to sacrifice all his prospects rather than disavow a tittle of his creed. Had that creed been one of greater significance, disciples would have revered him as a worthy martyr, and adversaries regarded him as dangerous in proportion to his virtue. Unluckily it was a creed untenable by any man of sound intellect." As we have seen, many Unitarians did revere Whiston as a worthy leader of a worthy cause.

99 *Encyclopedia Brittanica,* 11th ed. (London, 1910–11), s.v. "Whiston, William."

100 Farrell, *Whiston,* p. 35.

101 Ibid., p. 34.

102 Velikovsky acknowledges Whiston's influence on his controversial reconstruction of world history in his *Worlds in Collision* (London: Abacus, 1974), p. 53n and p. 316n. This work first appeared in 1950. Velikovsky is glad for Whiston's support of his theory of a cometary impact upon the earth in ancient times and of the consequences of this impact upon the length of the year.

103 See the most recent edition, *Josephus: Complete Works,* ed. William Sanfor LaSor, trans. William Whiston (Grand Rapids, Mich.: Kregel, 1981).

104 Benjamin Hoadley, "To His Holiness Clement XI," in Sir Richard Steele, *An Account of the State of the Roman-Catholick Religion throughout the World* (London, 1715), pp. xiii–xv. This letter is signed by Steele, but Hoadley is identified as the genuine author by George A. Aitken, *The Life of Richard Steele,* 2 vols. (London, 1889), 2:243–4.

2. WHISTON, THE BURNET CONTROVERSY, AND NEWTONIAN BIBLICAL INTERPRETATION

1 It was not explicitly evident to Descartes's English readers that the Cartesian mechanism must eventually pose the problem of how to interpret Mosaic history. Henry More's initial wild enthusiasm for the Cartesian system, for example, served to stimulate interest among the many young scholars he influenced. But in a letter to Boyle in 1665 and later in the *Divine Dialogues* in 1668, More renounced Cartesianism and placed Descartes with the atheistic Hobbists and "Epicureans" who attempted to explain all phenomena by purely mechanical causes. See C. Webster, "Henry More and Descartes: Some New Sources," *British Journal for the History of Sciences* 4, no. 16 (1969):359–77. For a brief catalog of Descartes's mechanistic explanation of cosmology (summarized below),

see Descartes, *Principia Philosophiae,* in *Oeuvres de Descartes,* ed. C. Adam and P. Tannery (Paris, 1897–1910), 12 vols. The key points are: Cosmogenic speculation is only hypothetical (8:99). Descartes's "opinions" nevertheless submitted to the authority of the church (329). Material substance is identical with pure extension, and pure extension is identical with space: Therefore, there cannot be a vacuum, for wherever there is space there is extension or matter. God creates this material plenum, the laws that govern its motion, and endows it with a certain amount of motion (49–51). The job is to construct through geometry a complete and certain physics of matter moving in space (151). Primordial matter is originally homogeneous in composition but not in size and apportioned equally around various localities in the plenum. Each of these localities has a center point around which the particles of undifferentiated matter, homogeneous matter, circulate. These whirlpools are called ***vortexes*** or ***tourbillons.*** Through friction with other particles, the corners of the larger bits of matter are eroded, and the plenum is gradually filled with infinitesimal dust particles moving at high speed. These dust particles, which are luminescent, constitute Descartes's first element. The eroded blocks of matter become spherical globes in the erosion process and are his second element. Matter is also ground up into a size and shape intermediate between the fast-moving, luminescent, infinitesimal dust particles and the large spheres. This is Descartes's third element (103–4). The luminescent, infinitesimal particles of the first element are drawn to the center of the vortex, where they cohere to form stars. The large globes of the second element are gradually driven to the periphery of the vortex, where they become the heavenly bodies, reflecting light from the central star. The third element, the grossest and most opaque, becomes the crust of planets and comets, through the mechanism of sunspots (104–8). Accretions of the third, opaque element on the surface of the glowing central sun ultimately cause sunspots which lead to a variety of changes from a variety of causes (135–8, 147–8). The swirling first-element particles might engulf these accretions, the accretions might break off the surface and be hurled back into the plenum, or they might grow like cancers with additional conglomerations of the third element, spreading to cover the entire surface of the primary matter of the central star with a thin crust. In this case, the light of the luminescent central body would be snuffed out, and less pressure would be exerted throughout the vortex (151). The elements of the vortex would slow down and be absorbed into neighboring vortexes. The now darkened central star would either be absorbed into the neighboring vortex and would become, according to its degree of solidarity, either a planet in orbit around the central sun of the new home vortex or comets wandering from vortex to vortex (156–7). According to Descartes's hypothesis, the earth was a former star captured by the solar vortex after the earth's fiery, luminescent core of the first element of matter had been covered over with various gradations of the opaque, inert third element that formed various levels of the earth's crust around the still simmering central core of the first element.

2 Blaise Pascal, *Pensées,* Pensée 77, in *Oeuvres de Blaise Pascal,* ed. Leon Brunschvicg (Vaduz, Liechtenstein: Kraus Reprint, 1965), 12:98. "I cannot forgive Descartes. In all his philosophy he would have been quite willing to dispense with God. But he had to make Him give a fillip to set the world in motion; beyond this, he has no further need of God." *Pascal's Pensées,* trans. W. F. Trotter (New York: E. P. Dutton, 1958), Pensée 77, p. 23.

3 Edward Stillingfleet, *Origines Sacrae, Or a Rational Account of the Grounds of Christian Faith, As to the Truth and Divine Authority of the Scriptures, And the matters therein contained* (London, 1662), pp. 468–9. In the posthumously published fragments of the *Origines Sacrae* written in 1697, the bishop declares an interest in the adequacy of Descartes's hypothesis that "matter being thus put into motion [by God], can produce the *Phenomena* of the World, without any further interposition of Providence; than only to preserve the motion of Matter." Cited in Robert Todd Carroll, *The Common-Sense Philosophy of Religion: Bishop Edward Stillingfleet, 1635–1699* (The Hague: Martinus Nijhoff, 1975), pp. 104–5. Carroll believes that the shift that Stillingfleet makes in asking whether the account given by Descartes is adequate or true represents "a shift in Stillingfleet's perception of the popularity of mechanistic theories in his own day" (p. 105). I would point out that the fragment expressing this new concern dates from the year after Whiston's *New Theory* was published. Unfortunately, Stillingfleet died before he could begin his reconsideration of the adequacy of celestial vortexes as an account of God's actions.

4 See D. C. Kubrin, "Providence and the Mechanical Philosophy: The Creation and the Dissolution of the World in Newtonian Thought" (Ph.D. diss., Cornell University, 1968), p. 92. Kubrin's work is the most helpful modern interpretation of Burnet. Other valuable perspectives have been provided by Marjorie Hope Nicolson, who shows Burnet's place in the development of Romantic aesthetics, in *Mountain Gloom and Mountain Glory: The Development of the Aesthetics of the Infinite* (Ithaca: Cornell University Press, 1959); Ernest Lee Tuveson who has placed Burnet in the context of the relationship of the idea of the Christian millennium to the idea of progress, in *Millennium and Utopia: A Study in the Background of the Idea of Progress* (New York: Harper Torchbooks, 1964); Francis Haber, who has located Burnet's place in the controversy over the age of the earth, in *The Age of the World: Moses to Darwin* (Baltimore: Johns Hopkins University Press, 1959); Don Cameron Allen, who sees Burnet, rightly, as the harbinger of a new Christian rationalism, in *The Legend of Noah: Renaissance Rationalism in Art, Science and Letters* (Urbana: University of Illinois Press, 1949); John C. Greene, who examines Burnet as one among many who seek scientific explanations for natural phenomena in conjunction with saving the "great doctrines" of revelation and creation, in *The Death of Adam: Evolution and Its Impact on Western Thought* (Ames: Iowa State University Press, 1959); Michael Macklem, who is concerned with the way such theorists as Burnet relate their developing concept of natural law to moral law, in *The Anatomy of the World: Relations between Natural and Moral Law from Donne to Pope* (Minneapolis: University of Minnesota Press, 1958); Basil Willey, who, like Kubrin, interprets Burnet as demonstrating God's providence over even this "wreck" of a world, in *The Eighteenth Century Background: Studies on the Idea of Nature in the Thought of the Period* (Harmondsworth, England: Penguin Books, 1972; 1st ed., 1940); by Katharine Brownell Collier, who gives a thorough digest of Burnet's particular theoretical speculations along with those of several other cosmogonists, in *Cosmogonies of Our Fathers: Some Theories of the Seventeenth and Eighteenth Centuries* (New York: Columbia University Press, 1934); and, more recently, by Jacques Roger, who convincingly argues that Burnet paved the way for an entirely historical interpretation of nature by his attempt to conjoin Cartesian cosmogony with the historical ac-

count of earth history contained in the Bible, in "La Théorie de la terre au XVIIe siècle," *Revue d'histoire des sciences* 26, no. 1 (1973):23–48.

5 Thomas Burnet, *Archaeologiae Philosophicae: Sive Doctrina de Rerum Originibus* (London, 1692). The seventh and eighth chapters of Burnet's book, translated by Henry Brown, were printed in Charles Blount, *Oracles of Reason* (London, 1693), from which this citation is taken, pp. 63–4.

6 Thomas Burnet, *The Sacred Theory of the Earth,* ed. Basil Willey (Carbondale: University of Southern Illinois Press, 1965), p. 93. This edition follows the London edition of 1691.

7 Ibid., p. 223.

8 Ibid., pp. 88–89.

9 Ibid.

10 Ibid., pp. 34, 58, 89.

11 Ibid., pp. 68, 70–1.

12 Ibid., pp. 151–8. In Burnet's theory, the antediluvian earth of the golden age was a perfect sphere, always equidistant from the sun. It rotated on its axis perpendicularly to the plane of the ecliptic. Because the axis always remained perpendicular, there was a true paradise that remained in "*perpetual Serenity and Perpetual Aequinox*" (ibid., p. 137). Because its axis always remained perpendicular to the plane of the ecliptic, the pristine earth was divided into zones – the "Frigid Zones" nearest the poles and the "Torrid Zone" around the equator. Between both poles and the equator were two "Temperate Zones" (ibid., p. 170) that, as a result of God's generally provident design, were eminently suitable to human habitation and cultivation. The "serenity" of the atmosphere, the ideal weather, and the fertility of the soil combined to produce extreme good health and longevity in all living things. But this perpetual summer gradually dried the exterior crust. The heat also rarified the watery vapors immediately beneath the crust, which built up pressure. And at length, "these preparations in Nature being made on either side, the force of the Vapours increas'd and the walls weaken'd, which should have kept them in, when the appointed time was come, that All-wise Providence had design'd for the punishment of a sinful World, the whole fabrick brake, and the frame of the Earth was torn in pieces" (ibid., p. 68). A new world appeared as the floodwaters subsided and ran back into the nooks and crannies of the uptilted slabs of crust, a world divided into our present earth and seas. Furthermore, as Burnet spun out his theory, the earth's center of gravity shifted when the ballast of the crust was so greatly disturbed. This shift tilted the earth's axis with respect to the plane of the ecliptic. The postdiluvian "skew-posture" destroyed the "more easie and regular disposition" of the sun, thus eliminating the perpetual equinox and introducing "that inequality of Seasons which hath since obtain'd" (ibid., p. 147). The harshness of the newly introduced seasons reduced the average life span from the six to nine hundred years common in the antediluvian world to the postdiluvian sixty years (ibid., pp. 151–8). "And so Divine Providence, having prepar'd Nature for so great a change, at one stroke dissolv'd the frame of the old World, and made us a new one out of its ruines, which we now inhabit since the Deluge" (ibid., pp. 70–1). Burnet's speculations about the stages of the earth's development, in the case of the cause of the Deluge and its after-effects, is exactly conformable, he claims, to the account in Scripture. The God of general providence who set it all up beforehand and the God revealed in revelation are exactly the same.

13 Ibid., p. 72.

14 Whiston, *A Vindication of the New Theory of the Earth from the Exceptions of Mr. John Keill, and others* (London, 1698), p. 2.

15 In his *Review of the Theory of the Earth, and of its Proofs, Especially in Reference to Scripture* (London, 1690), reprinted in the Willey edition of *Sacred Theory of the Earth,* pp. 381–412, Burnet repeats his argument for the radical differences between the antediluvian and postdiluvian earth. He concludes that only his theory agrees with Moses' history of both the extent of the Flood (Moses and Burnet claim that it covered the entire earth) and the mechanical causes (a rupture of the crust of the Cartesian third element, and a heavy rainfall). He criticizes in particular those theorists, such as Isaac La Peyrère, who in his notorious *Men Before Adam* (London, 1656), Book 4, claims that the Flood was a local event confined to Noah's immediate environment. Burnet also criticizes those who postulate the miraculous creation of the floodwaters *ex nihilo* and then their miraculous annihilation. Burnet claims that "both these explications you see, (and I know more of note that are not obnoxious to the same exceptions) differ from *Moses* in the substance, or in one of the two substantial points. . . . The first changeth the Flood into a kind of national inundation, and the second assigns other causes of it than *Moses* had assigned" (ibid., p. 407).

16 Ibid.

17 Ibid., p. 408.

18 Ibid., pp. 407–8.

19 Charles Blount, *Miracles, no violations of the laws of nature* (London, 1682), pp. 3–4. Spinoza's *Tractatus Theologico-Politicus* was published in 1677 in Rotterdam and formed the basis for Blount's pamphlet. In one of the replies to Blount, Sir Thomas Browne's *Miracles, works above and contrary to nature: or, an answer to a late translation out of Spinoza's Tractatus Theologico-Politicus* (London, 1683), Browne notes on p. 2 that in addition to plagiarizing from Spinoza, Blount also copied part of his preface from Thomas Burnet. See J. A. Redwood, "Charles Blount (1654–93), Deism, and English Free Thought," *Journal of the History of Ideas* 35, no. 3 (July–Sept. 1974):494–5. Blount had spotted the "freethinking" applications of Burnet's ideas quite early.

20 Blount, *Miracles,* pp. 30–1.

21 Charles Blount, ed., *The First Two Books of Philostratus, concerning the life of Apollonius Tyraneus* (London, 1680), pp. 32–3.

22 Burnet, *Archaeologiae Philosophicae,* Chapter 7. Quoted from Brown's translation in Blount, *The Oracles of Reason* (London, 1693), p. 75. Burnet notes many rational problems with the Mosaic creation story which indicate that "Moses has followed the popular System; that which pleases the People" rather than a rational, philosophical foundation: (1) Moses makes the earth the center "of the whole Machine" (p. 54). (2) Moses speaks as if the chaos "filled and possessed" the whole universe, which is impossible, because stars are "fiery bodies" and must precede the "Foundation of the Earth" (pp. 54–8). (3) Angels, too, must have existed before the establishment of the earth, or else we must suppose the Supreme Being no sooner had taken "his hand from off them, but they immediately fell headlong to destruction." Not only did they exist before the earth; the very fact of their fall indicates an admixture of matter, or else why did they fall? Hence, matter existed prior to the chaos of Moses (p. 59). (4) Descartes's hypothesis saves the phenomena of new stars, sunspots, and comets

(p. 63). (5) Moses refers to water from "above the firmament," whereas we have never seen such a thing. Again, because the vulgar do not sufficiently understand the natural generation of rain from the condensation of water vapors, they prefer to think of it as a miraculous act of special providence, "sent down from Heaven by a Divine Impulse" (p. 68).

23 Ibid., Chapters 15–16 passim, pp. 12–14, 16–17, 193.

24 The aim of deism, according to John Leland in 1754, was simply "to set aside revealed religion." Blount's work, as well as Burnet's, obviously contributed materially to this end. See Leland, *A View of the Principal Deistical Writers that have Appeared in England in the last and Present Century* (London, 1754), Preface, p. iii.

25 W. King, *Works* (London, 1776). Cited in *DNB*, s.v. "Burnet, Thomas."

26 See Blount, *Oracles of Reason*, pp. 1–19.

27 John Oldmixon, *History of England During the Reigns of King William and Queen Anne, King George I., being the Sequel to the Reigns of the Stuarts . . .* (London, 1735), pp. 95–6. Cited in Kubrin, "Providence and the Mechanical Philosophy," pp. 145–6, and DNB s.v. "Burnet, Thomas." Kubrin illuminates the connection between Blount and Burnet (pp. 146–7), as does John Redwood, *Reason, Ridicule and Religion: The Age of Enlightenment in England* (London: Thames & Hudson, 1976), pp. 122–3.

28 Burnet, "Preface to the Reader," in *Sacred Theory of the Earth*, p. 15.

29 Whiston, *A New Theory of the Earth, From its Original, to the Consummation of all Things, Where in the Creation of the World in Six Days, the Universal Deluge, And the General Conflagration, As laid down in the Holy Scriptures, Are Shewn to be perfectly agreeable to Reason and Philosophy. With a large Introductory Discourse concerning the Genuine Nature, Stile, and Extent of the Mosaick History of the Creation* (London, 1696). The text cited is from p. 2 of the separately paginated "Introductory Discourse."

30 Ibid., p. 64.

31 Ibid., pp. 2, 65–6.

32 Ibid., p. 64.

33 Ibid., p. 3.

34 Ibid.

35 Ibid.

36 Henry G. Van Leeuwen, *The Problem of Certainty in English Thought, 1630–1690* (The Hague: Martinus Nijhoff, 1963).

37 Whiston, *New Theory*, pp. 127–8.

38 Whiston cites Locke, "the best of our Metaphysicians," to show that the Being of God is an a priori "First Notion" (ibid., p. 128); Locke, *Essay Concerning Human Understanding*, Book 4, Chapter 10.

39 Whiston, *New Theory*, p. 128.

40 Ibid.

41 Ibid., p. 129.

42 Ibid., p. 131.

43 Ibid., p. 128.

44 Ibid., p. 132.

45 Whiston, "Introductory Discourse," p. 73. Whiston, secure in his faith that everything originates from the supernatural deity, argues here that there are

many reasons why this may not appear to be the case. Spinoza utilizes the same arguments about the limits of human reason that often prevent us from seeing that everything originates from nature (*sive Deus*). Cf. *Tractatus Theologico-Politicus,* Chapter 6. Cf. also John Locke's view that "reason is the proper judge; and revelation, though it may, consenting with, confirm its dictates, yet cannot in such cases invalidate its decrees" (*Essay Concerning Human Understanding,* ed. A. Fraser [New York, 1959], p. 423). Whiston, likewise, is adamant that "the Measure of our present knowledge ought not to be esteem'd the κριτήριον, or Test of Truth; or to be oppos'd to the Accounts receiv'd from profane Antiquity much less to the inspir'd writings" (*New Theory*, p. 379).

46 Whiston, "Introductory Discourse," pp. 4–7; cf. *New Theory,* pp. 222–3.

47 Whiston, "Introductory Discourse," p. 56. Though Whiston urges extreme caution in "arguing from Man to God" (ibid., p. 49), he asks, "Will a wise Builder bestow twice as much time in decking and adorning one By-closet of inferior use, and that only to some of the meanest Servants too; as of the Royal Palace, with all its stately Rooms and Apartments, intended for the King himself, and his Courtiers? Should we hear of such strange Actions, and disproportionate Procedure among Men, we should not be able to induce our selves to give credit there to" (ibid., p. 57).

48 Ibid., p. 30.

49 *New Theory,* pp. 74–5.

50 Ibid., pp. 47–8.

51 Ibid., p. 214.

52 Ibid., pp. 81–2.

53 Ibid., pp. 79–104. Whiston hypothesizes that it took the earth in its primitive, antediluvian state more time to revolve around the sun than it did after the Flood or now (p. 134).

54 Ibid., p. 232.

55 Ibid., pp. 238–9.

56 Ibid., pp. 241–2.

57 Burnet's earth with its mantle of crust stretched completely over the Abyss simply ignores Moses. When Whiston describes the production of mountains, plains, and valleys on the second day (year) of creation and not, like Burnet, at the time of the Flood, he argues that mountains are less dense and thus sink less deeply into the heavy fluids of the core, because: (1) mountains are the principle sources of springs and therefore porous (he cites Woodward, *An Essay Toward a Natural History of the Earth* [London, 1695], Part 3, as corroboration on this point); (2) they are often volcanic and hence composed of light and rare sulphurs; and (3) they are especially subject to earthquakes and therefore "Hollow and Cavernous, Loose and Spungy in their inward parts" (ibid., pp. 76–9).

58 Ibid., p. 246–7.

59 Ibid., pp. 224–5.

60 Ibid., pp. 249–50. Whiston believes that other planets were formed in a similar manner to the earth: " 'Tis very reasonable to believe, that a *Planet* is a *Comet* form'd into a regular and lasting constitution." ("Introductory Discourse," p. 80).

61 Whiston, *New Theory,* pp. 251–2.

62 Ibid., p. 253.

63 "Introductory Discourse," p. 95.
64 Whiston, *New Theory,* pp. 277–9, 282–300.
65 Ibid., 2nd ed. (London, 1708), pp. 111, 345–6.
66 Whiston's views concerning the degree of eccentricity in the orbit of the earth after it ceased its cometary path of radical eccentricity underwent change. In the first edition of *A New Theory,* he supposes that after creation the earth's orbit was concentric. By the time of the publication of the second edition in 1708, he hypothesizes that the earth, at creation, had a moderately elliptical orbit. After the Fall, the collision of a comet with the earth altered the species of the ellipse, changing the orbit into a perfectly concentric one. Then, when another comet passed close by the earth at the time of the Deluge, the orbit again was altered, changing from a perfect concentric circle back to a moderate ellipse. This view seemingly forces Whiston into the paradoxical notion that after the Fall the earth's orbit was more perfect than before, a circle being more perfect than an ellipse. Whiston answers that an elliptical orbit, such as existed before the Fall when man lived in paradise only, provided "peculiarly for the happiness for that particular spot" (*New Theory,* 2nd ed., p. 115). Kubrin has pointed out this change and traces it to Whiston's reaction to the criticism of John Keill. See "Providence and the Mechanical Philosophy," p. 282.
67 Whiston, *New Theory,* 2nd ed. pp. 432–4. Emphasis added. Even though Whiston identifies the actual daily operation of the laws of nature as evidence of God's prescient general providence at the time of creation, he argues as a corollary that as such prescience is so far above human beings, it is most "Obvious," "Scriptural," and "Prudent" always "to suppose an immediate Exerting of a new Power in every new Turn in the World" and so to continue to pray (ibid., p. 436).
68 Ibid., pp. 367–70.
69 Ibid., pp. 370–82.
70 Ibid., pp. 40–2.
71 Ibid., pp. 144–213.
72 Whiston, *The Cause of the Deluge Demonstrated: Wherein it is proved that the famous Comet of A.D. 1680, came by the Earth at its Deluge, and was the Occasion of it. Being an Appendix to the New Theory of Earth. The Second Edition with Additions* (London, 1714).
73 Whiston, *New Theory,* 2nd ed., pp. 440–9.
74 Whiston, "Introductory Discourse," p. 95.
75 Whiston, *Memoirs of the Life and Writings of Mr. William Whiston . . . ,* 2nd ed., 2 vols. (London, 1753), 1:43. Unless otherwise indicated, all citations are to this edition.
76 Cf. *DNB,* s.v. "Whiston, William"; Katharine Brownell Collier, *Cosmogonies of our Fathers,* pp. 109–24; E. G. R. Taylor, "The English World-Makers of the Seventeenth Century and their Influence on the Earth Sciences," *Geographical Review* 38 (1948):111; Ernest Tuveson, "Swift and the World-Makers," *Journal of the History of Ideas* 11, no. 1 (January 1950):56; and Robert H. Hurlbutt, *Hume, Newton, and the Design Argument* (Lincoln: University of Nebraska Press, 1965), p. 36. Interpretations more sensitive to the importance of Whiston's method of biblical interpretation that still fall short of identifying this method as a legitimate facet of Newtonianism are Kubrin, "Providence and the

Mechanical Philosophy," pp. 268–72; John C. Greene, *The Death of Adam*, Chapter 2; and Frank E. Manuel, *Isaac Newton, Historian* (Cambridge, Mass.: Belknap Press of Harvard University Press, 1963), p. 144.

77 Roy Porter, *The Making of Geology: Earth Science in Britain, 1660–1815* (Cambridge: Cambridge University Press, 1977), pp. 72–9. Newton's view of matter as passive unless acted upon by some outside force contrasts with the supremely hylozoic matter of Descartes. Thus the Cartesian Thomas Burnet points out that the one general feature about the creation of nature that the Mosaic account of creation gives with scientific accuracy, and shares with the Cartesian theory, is that "both suppose the Chaos to have been the matter out of which the World was Built" and that it was at first "inanimate, and then afterwards animated." See *Archaelogiae Philosophicae*, in Blount, *Oracles of Reason*, pp. 52–3. Whiston, as a good Newtonian, cites Richard Bentley and sets down as corollaries to the "Lemmata" of Book I of his *New Theory* that matter is entirely passive in its very essence. Thus gravity is yet another demonstration of divine providence: "This universal force of Gravitation being so plainly above, besides, and contrary to the Nature of Matter . . . must be the Effect of a Divine Power and Efficacy, which governs the whole World, and which is absolutely necesssary to its Preservation" (p. 6). Kubrin, "Providence and the Mechanical Philosophy," p. 260, has drawn attention to this aspect of Whiston's Newtonianism as well.

78 Whiston relies heavily on Newton's second rule of reasoning in philosophy, which states that "to the same natural effects we must, as far as possible, assign the same causes" (*Mathematical Principles of Natural Philosophy*, trans. Andrew Motte [1729], rev. Florian Cajori [Berkeley: University of California Press, 1934], p. 398). Cf. "Introductory Discourse," where Whiston states that "every unbyass'd Mind would easily allow, that like Effects had like Causes; and that Bodies of the same general Nature, Uses, and Motions, were to be deriv'd from the same Originals; and consequently, that the Sun and the fixed Stars had one, as the Earth, and the other Planets another sort of Formation. If therefore any free Considerer found that one of the latter sort, that Planet which we Inhabit, was deriv'd from a Chaos; by a parity of Reason he would suppose, every one of the other to be so deriv'd also" (p. 40).

79 Burnet to Newton, January 13, 1680/1, in *The Correspondence of Isaac Newton*, vol. 2, *1676–1687*, ed. H. W. Turnbull (Cambridge: Cambridge University Press, 1960), p. 324.

80 Ibid., p. 323.

81 Ibid., p. 324.

82 Ibid., p. 327.

83 Ibid., p. 325.

84 Ibid., p. 324.

85 Ibid.

86 Ibid., p. 322. This letter of December 23, 1680, no longer exists. The passage is taken from a slightly longer text for Newton's letter quoted by Burnet in his reply of January 13, 1681.

87 Ibid., p. 325.

88 Newton to Burnet, January 1681, in Turnbull, *Correspondence of Newton*, 2:331, 333.

89 Ibid., p. 332.

90 Ibid., p. 334.

91 Ibid., p. 332. Newton's own interest in the most ancient records as the basis for history is well detailed by Manuel, *Newton, Historian,* and by J. E. McGuire and P. M. Rattansi, "Newton and the 'Pipes of Pan,'" *Notes and Records of the Royal Society of London* 21, no. 2 (Dec. 1966): 108–43.

92 Newton to Burnet, January 1681, in Turnbull, *Correspondence of Newton,* 2:331. Danton B. Sailor, "Moses and Atomism," in *Science and Religious Belief: A Selection of Recent Historical Studies,* ed. C. A. Russell (London, 1973), p. 17, notes this text and inaccurately observes that Burnet's work is part of the "popular turn-of-the-century purpose of reconciling the Mosaic account with Newtonian science." Sailor goes on to suggest, also mistakenly, in my view, that "Newton was able to concur, not only with its purpose (which Newton shared enthusiastically) but with most of its contents." For an analysis of the biological and chemical bases of such analogies, see Henry Guerlac, "Theological Voluntarism and biblical Analogies in Newton's Physical Thought," *Journal of the History of Ideas* 44, no. 2 (April–June 1983):227–8.

93 Newton to Burnet, January 1681, in Turnbull, *Correspondence of Newton,* 2:332.

94 Ibid.

95 Ibid., p. 333.

96 Ibid., p. 334.

97 Newton, *Mathematical Principles,* 2:541.

98 David Gregory, *The Elements of Physical & Geometrical Astronomy* (London, 1726), Book II, p. 853.

99 Edmond Halley, "Of the Cause of the Universal Deluge," and "Farther Thoughts on the Same Subject," *Philosophical Transactions of the Royal Society,* 33 (London, 1724–25):118–25. When these papers were printed in 1724, Halley explained that he had not published them, after reading them in 1694 to the Royal Society, lest "by some unguarded Expression" he "might incur the Censure of the Sacred Order." Halley probably utilized the cometary hypothesis before Whiston to explain the Flood, but he did not use it to explain the "chaos," diurnal motion, or the end of the world. The fact that he later proposed Whiston for membership in the Royal Society suggests that he was not upset by Whiston's "appropriation" of his hypothesis. Whiston even acknowledges Halley's work and explains how his book goes beyond it, in *A New Theory.* The nature of a cometary orbit, all "competent judges" of the new astronomy agree, provides the perfect naturalistic explanation of the cause of a deluge: "But that it really did so at the time specified, is what I am now to prove. 'Tis true, when upon a meer Supposition of such a passing by of a Comet, I had in my own mind observ'd the Phaenomena relating to the Deluge to answer to admiration, I was not a little surpriz'd, and pleas'd at such a Discovery." This "meer Supposition" I take to refer to Halley's cometary hypothesis, which set Whiston to thinking how ideal a scientific explanation such a hypothesis was and set him off on the search for more corroborating evidence, which he found and published in *A New Theory,* "to clear the Holy Scriptures from the Imputations of ill-despised Men, and to demonstrate the Account of the Deluge to be in *every part* neither impossible nor unphilosophical" (pp. 126–7; emphasis added). Cf. Marjorie Nicolson and G. S. Rousseau, *"This Long Disease, My Life," Alexander Pope and the Sciences* (Princeton: Princeton University Press, 1968), p. 140n.

100 Whiston, "Reason and Philosophy no Enemies of Faith," in *Sermons and Essays upon Several Subjects* (London, 1709), p. 210.
101 Whiston, *Astronomical Principles of Religion, Natural and Reveal'd* (London, 1717), p. 242.
102 Ibid., p. 133.
103 Newton to Bentley, December 10, 1692, in Turnbull, *Correspondence of Newton*, 2:233.
104 Newton's "method of analysis" is the same as Whiston's fourth "Probation of knowledge." His initial, very brief statement of this method is in the Preface to the first edition of the *Principia*, but it is refined and expanded in the third edition of the *Opticks* (London, 1718), where Newton states that "as in Mathematicks, so in Natural Philosophy, the investigation of difficult Things by the Method of Analysis, ought ever to precede the Method of Composition. This Analysis consists of making Experiments and Observations, and in drawing general Conclusions from them by Induction, and admitting of no Objections against the Conclusions, but such as are taken from Experiments, or other certain Truths. For Hypotheses are not to be regarded in experimental Philosophy. And although the arguing from Experiments and Observations by Induction be no Demonstration of general Conclusions; yet it is the best way of arguing which the Nature of Things admits of. . . . By this way of Analysis we may proceed from Compounds to Ingredients; and from Motions, to the Forces producing them; and, in general, from Effects to their Causes, and from particular Causes to more general ones, till the Argument end in the most general" (Query 31, p. 380). Much of the query is devoted to a statement of the design argument.
105 Newton to Bentley, December 10, 1692, in Turnbull, *Correspondence of Newton*, 2:235. For a thorough exposition of Newton's version of the design argument and his theism in general, see Hurlbutt, *Hume, Newton, and the Design Argument*, pp. 3–26, and E. W. Strong, "Newton and God," *Journal of the History of Ideas* 13, no. 2 (April 1952), pp. 147–67.
106 Newton, *Principles*, p. 544.
107 E. A. Burtt has argued that for Newton these first principles also comprised "a body of absolutely certain truth about the doings of the physical world." See *The Metaphysical Foundations of Modern Science* (New York, 1927), p. 223. Van Leeuwen, in *Problem of Certainty in English Thought*, p. 119, has pointed out that Newton was not so absolute as Burtt claims. Though Newton was more circumspect in ascribing certainty to his system, Whiston's view that these principles were indeed certain was more widespread, and largely accounted for Newton's universally esteemed reputation. See, for example, J. T. Desaguliers, *The Newtonian System of the World, the Best Model of Government: An Allegorical Poem. With a plain and intelligible Account of the System of the World, by Way of Annotations* (London, 1728), pp. 21–2; Dedication by J. T. Desaguliers in W. J. S. Van's Gravesande, *Mathematical Elements of Natural Philosophy Confirmed by Experiments, or an Introduction to Sir Isaac Newton's Philosophy*, ed. and trans. J. T. Desaguliers (London, 1720), p. ii; William Derham, in the first edition of his *Astro-Theology: Or a Demonstration of the Being and Attributes of God, from a Survey of the Heavens* (London, 1715), p. 148, accords the hypothesis of gravity only "highly probable" status, p. 148. By the second edition, however, (London, 1715), he states that "upon this highly probable, I may say Physically certain, Theory of Gravity acting in the Motion of the Globes, we

have another exquisite Nicety in the Works of Creation" (p. 158). Roger Cotes, in the Preface to the second edition of Newton's *Principia,* edges in the direction of declaring Newton's hypotheses to be certain. Richard Bentley, in his presentation of the design argument in his Boyle Lectures of 1692, describes the scientific principles of Newton's *Principia* as "Truth solidly established." See his lectures in *A Defence of Natural and Revealed Religion: Being a Collection of the Sermons Preached at the Lecture founded by the Honourable Robert Boyle, Esq., (From the Year 1691 to the Year 1732),* 3 vols., ed. S. Letsome and J. Nicholl, *Dr. Bentley's Eight Sermons,* Sermon 7, 1:65. Desaguliers, Derham, Cotes, and Bentley all argue for the certain necessity of an "almighty architect" on the foundation of these sure and certain principles.

108 Whiston, *Astronomical Principles,* p. 40.

109 Ibid., p. 82.

110 Ibid., p. 83.

111 Ibid., p. 106. Whiston later defended the analogical method of reasoning: "And if there be any Deductions of Human Reason which are easier and more obvious than the rest, this way of Arguing, which we have already used, from the *House* to the *Architect;* from the *Clock* to the *Clockmaker;* from the *Ship* to the *Shipbuilder;* and from a *noble, large,* well-contriv'd and well-proportion'd, and most beautiful House, or Clock, or Ship, to the *excellent* Architect, the *skilful* Clockmaker, the sagacious Shipbuilder; this is such clear natural obvious, sure Reasoning, that we even at first make use of it in Childhood, and find it as clear, natural, obvious, and sure in our elder Age; without occasion for a tutor to instruct us in it at first, or for a Logician to improve us in it afterward" (ibid., p. 255). Totally forgotten was Whiston's own warning, in *A New Theory,* to beware of any argument that proceeded from man to God (see note 47).

112 Ibid., pp. 111–12.

113 Whiston, *New Theory,* 2nd ed., p. 284.

114 This text is from David Gregory's memoranda entitled "Annotations Physical, Mathematical and Theological from Newton, 5, 6, 7 May 1694," in Turnbull, *Correspondence of Newton,* Vol. 3, *1688–1694,* p. 336.

115 Whiston, *New Theory,* 2nd ed., pp. 6–7.

116 Whiston, *Astronomical Principles,* p. 112.

117 Ibid., pp. 134, 136.

118 Ibid., p. 134.

119 Cf. the first with the second paragraphs of the "General Scholium" in Newton, *Mathematical Principles,* 2:543.

120 Whiston, *Astronomical Principles,* pp. 136–7.

121 Ibid., pp. xxv–xxvi. The theme of using the method of the law courts in weighing historical evidence of human testimony while recognizing the fallibility in all such judgments is one of Whiston's favorites. He refers elsewhere to judges as "the most impartial and successful *Judges of Controversy* now in the World." Cf. *A Supplement to the Literal Accomplishment of Scripture Prophecies* (London, 1725), pp. 5–6. On the historical development of the legal theory concerning the degree of certainty obtainable in court, see Theodore Waldman, "Origins of the Legal Doctrine of Reasonable Doubt," *Journal of the History of Ideas* 20, no. 3 (June–Sept. 1959):299–316.

122 Whiston, *Astronomical Principles,* pp. xxvii, 134.

123 Michael Hunter, *Science and Society in Restoration England* (Cambridge: Cambridge University Press, 1981), p. 185. Cf. Jacob, *The Newtonians and the English Revolution, 1689–1720* (Ithaca: Cornell University Press, 1976), Chapters 4–5, and Kubrin, "Providence and the Mechanical Philosophy," pp. 219–33, 319–37, Chapters 10–11.

124 The best analysis of Keill's metaphysical Newtonianism is in E. W. Strong, "Newtonian Explications of Natural Philosophy," *Journal of the History of Ideas* 17, no. 1 (January 1957):49–83. Soon after his return to Oxford in 1712 as Savilian Professor of Astronomy, he began teaching Newtonian natural philosophy. The Newtonian disciple J. T. Desaguliers, who succeeded Keill as lecturer in experimental philosophy at Hart Hall, Oxford, credits Keill with being the first to teach Newtonian natural philosophy by experiments in a mathematical manner. Keill also played an important part in Newton's controversy with Leibniz over the invention of calculus.

125 John Keill, *An Examination of Dr. Burnet's Theory of the Earth: with Some Remarks on Mr. Whiston's New Theory of the Earth. Also an Examination of the Reflections on the Theory of the Earth; And a Defence of the Remarks on Mr. Whiston's New Theory,* 2nd ed. rev. (London, 1734), p. 140. Emphasis added.

126 Ibid., p. 107.

127 Keill, *A Defence of the Remarks Made on Mr. Whiston's New Theory,* p. 347.

128 In his hypothetical explanation of creation, Whiston reckons the creation of man a specially provident, miraculous interposition of God into the generally provident natural order. In his *Historical Memoirs of Dr. Samuel Clarke* (London, 1730), Whiston notes that Dr. Sykes persuaded Clarke to leave out of future editions of his Boyle Lectures Phlegon's account of a solar eclipse and a great earthquake at the time of Christ's Passion that are also mentioned by the Evangelists. Sykes's reasoning was that Phlegon's account was only a "supposal" and not capable of certainty. Whiston, however, made some calculations to see whether an eclipse could have occurred naturally at that time and found that no such natural eclipse would have been possible, "but only that *Supernatural* one at the Passion, which exactly agreed to it" (ibid., p. 148). Whiston later wrote a tract entitled *The Testimony of Phlegon Vindicated: or, An Account of the great Darkness and Earthquake at our Savior's Passion, described by Phlegon. Including all the Testimonies, both Heathen and Christian, in the very Words of the Original Authors, during the first Six Centuries of Christianity* (London, 1732).

129 Kubrin, "Providence and the Mechanical Philosophy," p. 318.

3. WHISTON'S NEWTONIAN ARGUMENT FROM PROPHECY; DIVINE PROVIDENCE; AND THE CRITICISM OF ANTHONY COLLINS

1 Margaret C. Jacob, *The Newtonians and the English Revolution, 1689–1720* (Ithaca: Cornell University Press, 1976), p. 270. For her argument that the Boyle lecturers, especially Richard Bentley and Samuel Clarke but including John Harris, William Derham, and William Whiston, adapted and refined Newtonian natural philosophy to serve as the "new underpinning of liberal Protestant social ideology" (p. 163), see Chapter 5.

2 Michael Hunter, *Science and Society in Restoration England* (Cambridge: Cambridge University Press, 1980), p. 186. Geoffrey Holmes also suggests, rather more caustically, that "heretical though the suggestion may seem to a historian of ideas, even the impact of the Boyle Lectures themselves *at the time* can be seriously exaggerated" ("Science, Reason, and Religion in the Age of Newton," *British Journal for the History of Science* 11, no. 38 [1978]:169).

3 Whiston, "Introductory Discourse," in *A New Theory of the Earth* (London, 1708), p. 95.

4 Jacob, *Newtonians and the English Revolution,* p. 269.

5 Newton to Bentley, December 10, 1692, in *The Correspondence of Isaac Newton,* 7 vols. (Cambridge: Cambridge University Press, 1959–77), vols. 1–3, ed. H. W. Turnbull, 3:233. Bentley's eighth and final Boyle Lecture was read on November 5, 1692.

6 Whiston, *Astronomical Principles of Religion, Natural and Reveal'd* (London, 1717), pp. 45–6.

7 Ibid., p. 243.

8 Ibid., pp. 242, 245.

9 Ibid., p. 242.

10 Ibid., pp. 243–4.

11 Cf. James O'Higgins, *Anthony Collins, the Man and His Work* (The Hague: Martinus Nijhoff, 1970), p. 174. Also, see Leslie Stephen, who shares Whiston's opinion that Woolston was unhinged, *History of English Thought in the Eighteenth Century,* 2 vols., (New York: Harcourt, Brace & World, 1962), 1:195. E. C. Mossner gives a good account of Woolston and his talent for fiendish ridicule without *odium theologicum* in *Bishop Butler and the Age of Reason* (New York: Macmillan, 1936), p. 75.

12 Whiston, *Memoirs of the Life and Writings of Mr. William Whiston . . .,* 2nd ed., 2 vols. (London, 1753), 1:197–201. Unless otherwise indicated, all citations are to this edition. The subsequent quotations in the next paragraphs are taken from these pages, and from pp. 236–38.

13 Stephen, *History of English Thought in the Eighteenth Century,* 2:138.

14 Whiston, *Memoirs,* 1:158.

15 Ibid., pp. 97, 99.

16 Ibid., p. 99.

17 John Leland, *A View of the Principal Deistical Writers that have Appeared in England in the Last and Present Century* (London, 1754), p. iii.

18 Whiston, *Historical Memoirs of Dr. Samuel Clarke* (London, 1730).

19 Whiston, *A Collection of Authentick Records Belonging to the Old and New Testament,* 2 vols. (London, 1727), 2:644.

20 Whiston, *Mr. Whiston's Account of the Exact Time When Miraculous Gifts Ceas'd in the Church* (London, 1728), p. 7.

21 Locke, *The Reasonableness of Christianity* (London, 1695), p. 55.

22 *Sir Isaac Newton: Theological Manuscripts,* ed. H. McLachlan (Liverpool: Liverpool University Press, 1950), p. 17.

23 Newton, *Observations upon the Prophecies of Daniel and the Apocalypse of St. John* (London, 1733), p. 25.

24 Joseph Mede, *"Epistles," being answers to divers Letters of Learned Men,* in *The Works of Joseph Mede,* ed. J. Worthington (London, 1672), p. 787.

25 Mede, "The Apostasy of Latter Times," in *Works*, p. 654.
26 Frank E. Manuel, *Isaac Newton: Historian* (Cambridge, Mass.: Harvard University Press, 1963), p. 164.
27 Newton, *Observations upon the Prophecies*, pp. 16–17.
28 The principle of reading a prophetic day as a modern year was accepted by Mede and all historians who interpreted the fulfillment of prophetic events, including Newton, Whiston, and the discoverer of logarithms, John Napier. See Napier, *A Plaine Discovery of the whole Revelation of St. John* (Edinburgh, 1593), p. 2.
29 Newton, *Observations upon the Prophecies*, p. 276.
30 Ibid., p. 25.
31 Ibid., pp. 130–1.
32 Joseph Mede, "*Remaines of Some Passages in the Apocalypse*," in *Works*, p. 603.
33 Newton, *Observations upon the Prophecies*, pp. 252–3. For more details on Whiston's and Newton's views concerning millennial prophecies, see Chapter 4, Section III.
34 Whiston, *Six Dissertations* (London, 1734), p. 270.
35 Henry Guerlac and Margaret C. Jacob, "Bentley, Newton, and Providence," *Journal of the History of Ideas* 30, no. 3 (July–Sept. 1969):307–18.
36 Cited in Guerlac and Jacob, "Bentley, Newton, and Providence," p. 317.
37 Samuel Pepys, *Private Correspondence and Miscellaneous Papers of Samuel Pepys*, ed. J. R. Tanner, 2 vols. (London: G. Bell & Sons, 1926), 1:51–2. Cited in Guerlac and Jacob, "Bentley, Newton, and Providence," pp. 317–18.
38 Guerlac and Jacob, "Bentley, Newton, and Providence," p. 318.
39 Whiston, *A Supplement to the Literal Accomplishment of Scripture Prophecies* (London, 1725), pp. 4–5.
40 Whiston, *Memoirs*, 1:98.
41 Newton, *Observations upon the Prophecies*, p. 252.
42 Whiston, *The Accomplishment of Scripture Prophecy* (London, 1708), p. 30.
43 Samuel Johnson, *A Dictionary of the English Language: in which the Words are deduced from their originals, and illustrated in their different significations by examples from the Lost writers*, 2 vols. (London, 1755), s.vv. "prophecy" and "to prophesy."
44 I am paraphrasing M. F. Wiles, "Origen as Biblical Scholar," in *The Cambridge History of the Bible*, vol. 1, *From the Beginnings to Jerome*, ed. P. R. Ackroyd and C. F. Evans (Cambridge: Cambridge University Press, 1976), p. 482.
45 Thomas's tenth article, in Question 1 of Part 1 to *The Summa Theologica*, is entitled "Whether in Holy Scripture a Word may Have Several Senses?" Thomas answers, with Augustine, that one scriptural passage may indeed have multiple levels of meaning: "Nevertheless nothing of Holy Scripture perishes because of this, since nothing necessary to faith is contained under the spiritual sense which is not elsewhere put forward clearly by the Scripture in its literal sense." See *Introduction to St. Thomas Aquinas*, ed. Anton C. Pegis (New York: Modern Library, 1948), p. 19.
46 Whiston, *Supplement*, p. 3.
47 Whiston, "Introductory Discourse," p. 95.
48 Whiston, *Accomplishment of Scripture Prophecy*, p. 13.
49 Ibid.
50 Whiston, *Supplement*, p. 5.

51 Ibid., p. 1.
52 Whiston, *Memoirs,* 1:191.
53 Ibid., p. 97.
54 Samuel Parker, *Censura Temporum,* 1, no. 1 (London, 1708), p. 19. Parker believes, however, that Whiston often speaks with too much assurance of the near approach of the "Millennial State" (ibid.).
55 Anthony Collins, *A Discourse of the Grounds and Reasons of the Christian Religion* (London, 1724), p. 51.
56 Whiston, *An Essay Towards Restoring the True Text of the Old Testament And for Vindicating the Citations made thence in the New Testament* (London, 1722), p. 1.
57 Ibid., p. 92.
58 Whiston, *Accomplishment of Scripture Prophecy,* p. 15; William Nicholls, *A Conference with a Theist* (London, 1696), 3:19.
59 Richard Simon, *Histoire critique du Vieux Testament* (Rotterdam, 1685), p. 495.
60 Whiston, *An Essay Towards Restoring the True Text,* pp. 115, 228.
61 Ibid., p. 232.
62 Collins, *Discourse,* p. 196.
63 Whiston, *Essay Towards Restoring the True Text,* p. 330.
64 Godfrey R. Driver, "Introduction to the Old Testament," in *The New English Bible* (Harmondsworth, U.K.: Penguin Books, 1974), p. xvi. Driver was a joint director of the Joint Committee on the New Translation of the Bible, a committee formed after World War II to translate the Bible into contemporary English. *The New English Bible* was first published in 1970 by Oxford University Press and Cambridge University Press.
65 Compare Whiston, *Essay Towards Restoring the True Text,* pp. 329, 333, and Driver, "Introduction to the Old Testament," pp. xvi–xvii.
66 Whiston, *Essay Towards Restoring the True Text,* p. 333.
67 Driver, "Introduction to the Old Testament," p. xv.
68 Whiston does cite other "dependable" sources not utilized by the translators of the *New English Bible,* e.g., Philo, Josephus, and writers from the hermetic tradition. See his "Appendix" to his *Essay.*
69 Driver, "Introduction to the Old Testament," p. xvii.
70 Whiston, *Essay Towards Restoring the True Text,* p. 335.
71 Collins, *Discourse,* p. 242.
72 Ibid., pp. 24, 28.
73 Ibid., p. 190. Cf. Simon, *Histoire critique du Vieux Testament,* p. 277.
74 Collins, *Discourse,* p. 193.
75 Collins, *Discourse* (London, 1724), p. 31.
76 Ibid., p. 43.
77 Ibid., p. 60.
78 Collins, *Scheme of Literal Prophecy Considered* (London, 1727), Preface.
79 Stephen, *History of English Thought in the Eighteenth Century,* 1:183.
80 Henry G. Van Leeuwen, *The Problem of Certainty in English Thought, 1630–1690* (The Hague: Martinus Nijhoff, 1963), pp. 5–6.
81 Whiston, *Supplement,* pp. 5–6.
82 Whiston, *Accomplishment of Scripture Prophecy,* p. 33.
83 James E. Force, "Hume and Johnson on Prophecy and Miracles: Historical Context," *Journal of the History of Ideas* 43, no. 3 (July–Sept. 1982):463–75.

84 Samuel Parker, *Censura Temporum,* vol. II (August 1709), pp. 613–14.

85 Whiston, *Memoirs,* 1:119–20; cf. Whiston. *Memoirs of Clarke* (London, 1730), p. 68. On Whiston and the French Prophets, see Hillel Schwartz, *The French Prophets: The History of a Millenarian Group in Eighteenth-Century England* (Berkeley: University of California Press, 1980), p. 37.

4. WHISTON'S NEWTONIAN BIBLICAL INTERPRETATION AND THE RAGE OF PARTY, RADICAL ARIANISM, AND MILLENNIAL EXPECTATIONS

1 For an account of the pioneering efforts of the first generation of the Royal Society in revealing the providential design and continual regulation and preservation of creation, see James E. Force, "Hume and the Relation of Science to Religion among Certain Members of the Royal Society," *Journal of the History of Ideas* 45, no. 4 (Oct.–Dec. 1984).

2 Margaret C. Jacob, *The Newtonians and the English Revolution: 1689–1720* (Ithaca: Cornell University Press, 1976), pp. 18, 51.

3 Ibid., p. 269.

4 Geoffrey Holmes, "Science, Reason, and Religion in the Age of Newton," *British Journal for the History of Science* 11, part 2, no. 38 (July 1978):168.

5 Whiston, *Memoirs of the Life and Writings of Mr. William Whiston . . . ,* 2nd ed., 2 vols. (London, 1753), 1:27. Unless otherwise indicated, all citations are to this edition.

6 Ibid., pp. 205, 258, 260. See also Chapter 1.

7 Holmes, "Science, Reason, and Religion," p. 168.

8 Whiston, *Memoirs,* p. 145.

9 Richard S. Westfall, *Never at Rest: A Biography of Isaac Newton* (Cambridge: Cambridge University Press, 1980), p. 651.

10 Richard G. Olson, "Tory–High Church Opposition to Science and Scientism in the Eighteenth Century: The Writings of John Arbuthnot, Jonathan Swift, and Samuel Johnson," a paper presented November 18, 1978, in a series of seminars held in 1978–9 at the William Andrews Clark Memorial Library at UCLA, entitled "Science, Technology, and Society in Postrevolutionary England." Professor Olson kindly provided me with a copy of this excellent article, which, along with the other seminar papers, is being published by the University of California Press.

11 Whiston, *Memoirs,* 1:208.

12 *Examiner,* IV. 41 (Oct. 26, 1714). Cited in *The Englishman,* ed. Rae Blanchard (Oxford: Clarendon Press, 1955), p. 426.

13 *Examiner,* V, 12 (Jan. 8, 1715). Cited in Blanchard, *The Englishman,* p. 426.

14 See Chapter 1 for the Scriblerian criticism of Whiston's longitude scheme and Chapter 5 for the attack on his "special" project of lectures predicting the millennium.

15 Whiston, *Memoirs,* 1:152.

16 See Larry Stewart, "Samuel Clarke, Newtonianism, and the Factions of Post-Revolutionary England," *Journal of the History of Ideas* 42, no. 1 (Jan.–March 1981):43–72; J. P. Kenyon, "The Revolution of 1688: Resistance and Contract," in *Historical Perspectives: Studies in English Thought and Society in Honour of J. H. Plumb,* ed. Neil McKendrick (London: Europa, 1974), pp. 43–69; Geoffrey Holmes, "Harley, St. John and the Death of the Tory Party," in *Britain after the Glorious Revolution,* ed. Geoffrey Holmes (London: Macmillan, 1969), pp. 216–

37; Gerald Straka, "The Final Phase of Divine Right Theory in England, 1688–1702," *English Historical Review* 77 (Oct. 1962):638–58; G. V. Bennett, "Conflict in the Church," in Holmes, *Britain after the Glorious Revolution,* pp. 155–75; Howard Erskine-Hill, "Alexander Pope: The Political Poet in His Time," *Eighteenth-Century Studies* 15, no. 2 (Winter 1981–82):123–48; James R. Jacob and Margaret C. Jacob, "The Anglican Origins of Modern Science: The Metaphysical Foundations of the Whig Constitution," *Isis* 71, no. 257 (June 1980):251–67; G. V. Bennett, "The Convocation of 1710: An Anglican Attempt at Counter-Revolution," in *Councils and Assemblies: Papers Read at the Eighth Summer Meeting and the Ninth Winter Meeting of the Ecclesiastical History Society,* ed. G. J. Cuming and Derek Baker (Cambridge: Cambridge University Press, 1971), pp. 311–19; Michael Hunter, *Science and Society in Restoration England* (Cambridge: Cambridge University Press, 1981); J. P. Kenyon, *Revolution Principles: The Politics of Party, 1689–1720* (Cambridge: Cambridge University Press, 1977); Geoffrey Holmes, *British Politics in the Age of Anne* (London: St. Martin's Press, 1967); Holmes, *The Trial of Dr. Sacheverell* (London: Eyre Methuen, 1973).

17 See Erskine-Hill, "Alexander Pope," pp. 124–6. Erskine-Hill, who sketches this contrast between the results of the standard interpretation and the newer understanding attained by the more recent work of the scholars cited in the preceding note, cites the following as sources for the older, "standard" interpretation traced in the text: Basil Williams, *The Whig Supremacy* (Oxford: Oxford University Press, 1939); J. H. Plumb, *Sir Robert Walpole,* vol. 1, *The Making of a Statesman,* and vol. 2, *The King's Minister* (London: Cresset, 1959, 1960); J. H. Plumb, *The Growth of Political Stability in England, 1675–1725* (London: Macmillan, 1967).

18 Benjamin Hoadley, *Works* (London, 1773), 2:177–8. Cited in J. P. Kenyon, "Revolution of 1688," p. 43.

19 Straka, "Final Phase," traces the contributions of these latitudinarian bishops in providing a rationale for the right to resist a king whose divine right to rule has been providentially withdrawn by God.

20 Holmes, "Science, Reason, and Religion in the Age of Newton," p. 168.

21 According to Holmes, in *Trial of Sacheverell,* p. 78, the Whigs were forced to vindicate the "Revolution principles" in the public spectacle of an impeachment proceeding because "at the end of more than a decade of political exertion in the Tory cause by the great bulk of the Anglican clergy, one of these parsons had preached a sermon in the heart of London which, on any reasonable construction, was seditious in implication and which was undoubtedly seditious in intent." Never before had a Tory gone so far in openly extolling the principle of passive obedience and in challenging Parliament's authority to pass such laws as the Toleration Act of 1695. The Whig government could not ignore such a direct threat to the legitimacy of the Crown and of Parliament.

22 Lord John Somers, *The Judgement of Whole Kingdoms and Nations, Concerning the Rights, Power, and Prerogative of Kings, and the Rights, Priviledges, and Properties of the People* (London, 1710), pars. 86–101. Somers also cites many instances illustrating this natural right drawn from Scripture (pars. 46–7, 84–5).

23 Whiston, *Memoirs,* 1:208–9.

24 Benjamin Hoadley, *The Original and Institution of Civil Government, Discuss'd, viz. I. An Examination of the Patriarchal Scheme of Government. II. A Defense*

of Mr. Hooker's Judgement . . . To which is added, A Large Answer to Dr. F. Atterbury's Charge of Rebellion: In which the Substance of his late Latin Sermon is produced, and fully examined (London, 1710), p. 101.

25 Whiston, *Memoirs,* 1:31–2.

26 *The Tryal of Dr. Henry Sacheverell, Before the House of Peers, for High Crimes and Misdemeanours; upon an Impeachment by the Knights, Citizens and Burgesses in Parliament Assembled, in the Name of themselves, and of all the Commons of Great Britain: Begun in Westminster Hall the 27th Day of February, 1709/10, and from thence continued by several Adjournments until the 23rd Day of March following* (London, 1710), p. 77. For all citations to *The Tryal,* I have drawn on the three hundred thirty-five–page edition listed in F. Madan, *A Bibliography of Dr. Henry Sacheverell* (Oxford: Printed by the Author, 1884), as "556," p. 23.

27 Whiston, *Memoirs,* 1:33.

28 *Tryal of Sacheverell,* p. 23.

29 Ibid., p. 73.

30 James Henry Monk, *The Life of Richard Bentley, D.D.*, 2 vols. (London, 1833), 1:286–90.

31 Caroline Robbins, *The Eighteenth Century Commonwealthman* (New York: Atheneum, 1968), p. 87.

32 For a description of the Newtonian effort to combat freethinking, crypto-Republican "Whiggism" – a tradition described by the authors as the Radical Enlightenment, in contrast to their notion of the latitudinarian, Anglican, Newtonian Enlightenment – see Jacob and Jacob, "Anglican Origins of Modern Science," passim, but esp. pp. 264–7. The Jacobs have been criticized for overemphasizing the importance of this tradition of free thought "at a time when heterodoxy was more diffuse." See Michael Hunter, *Science and Society in Restoration England,* p. 216, but I believe that the freethinking radical Enlightenment was, in fact, a serious opponent for the Newtonians such as Whiston who sought a middle ground between Jacobitism, on the one hand, and scoffing, freethinking, leveling deism on the other. See the record of Whiston's conversation with Newton about why Newton finally set forth his own version of the design argument, which is cited below in the text. See also the record of the nature of the meetings at Martin Folkes's house in 1721 by William Stukeley, in Chapter 5, Section I.

33 Whiston, *A Collection of Authentick Records Belonging to the Old and New Testament,* 2 vols. (London, 1728) 2:1073–4. (This is precisely the reason given by "Raphael Hythlodaeus" for not becoming a courtier in Part I of Thomas More's *Utopia.*) Whiston goes on to lament in this work the lack of a proper means for electing bishops (i.e., according to their Christian virtue) and bemoans the fact that in the modern church bishoprics are political plums to be dispensed to those whose loyalty can be relied upon or at least co-opted (pp. 1075–6; *Memoirs,* 1:235). Whiston's refusal of Jekyll's offer of a bishopric seems clearly related to his desire to maintain his integrity – an obsession of his described in Chapter 1 – and his fear of being corrupted by becoming dependent on a court for preferment, in the same manner as former close friends such as Lord Peter King and Bishop Benjamin Hoadley had.

34 Whiston, *Memoirs,* 1:259.

35 Whiston, *The Supposal* (London, 1712), reprinted in Whiston, *Scripture Politicks: Or an Impartial Account of the Origin and Measures of Government Ec-*

clesiastical and Civil, Taken out of the Old and New Testament (London, 1717), pp. 139–48.

36 Z. S. Fink, "Political Theory in *Gulliver's Travels*," *English Literary History* 14 (1947):157–8. Cf. Phillip Harth, "The Problem of Political Allegory in *Gulliver's Travels*," *Modern Philology* 73, No. 4, Pt. 2 (May 1976): S46, and Pat Rogers, "Swift and Bolingbroke on Faction," *Journal of British Studies* 9 (May 1970): 71–101.

37 See especially Whiston, *An Humble and Serious Address to the Princes and States of Europe, For the Admission, or at least open Toleration of the Christian Religion in their Dominions* (London, 1716). By "Christian Religion," of course, Whiston means primitive Christianity, i.e., Arianism.

38 Whiston, *Memoirs,* 1:27.

39 Straka, "Final Phase," pp. 648–52.

40 Whiston, *Scripture Politicks,* p. 18.

41 Ibid., pp. 23, 28, 31.

42 Ibid., p. 37.

43 Ibid., p. 46.

44 Ibid., p. 54.

45 Ibid., p. 94.

46 Ibid., p. 138.

47 Straka, "Final Phase," p. 658.

48 See John Lindsay, *A Short History of the Regal Succession: And the Rights of the several Kings recorded in the Holy Scriptures. With a Postscript occasion'd by Mr. Whiston's new Book of Scripture Politicks* (London, 1717). Lindsay, one of the last of the nonjurors, asserts on scriptural ground the necessity of passive obedience and of hereditary succession and vigorously denies Whiston's assertion that the Bible legitimizes the doctrine of providentially guided free election of monarchs. Lindsay quips that Whiston "seems to have studied other *Politicks,* more than those of the Holy *Scripture*" (p. 47).

49 Whiston, *Scripture Politicks,* p. 138.

50 Whiston, *An Historical Preface to Primitive Christianity reviv'd. With an Appendix Containing an Account of the Author's Persecution at, and Banishment from the University of Cambridge* (London, 1711), p. 17.

51 Whiston, *Memoirs,* 1st ed., 2 vols. (London, 1749), 1:27.

52 Whiston, *Sermons and Essays upon Several Subjects* (London, 1709), p. 242.

53 Ibid., pp. 247–8.

54 Whiston, *Sermons and Essays,* p. 245.

55 See Westfall, *Never at Rest,* pp. 310–19. As a result of a careful analysis of the Yehuda and Keynes manuscripts, Westfall makes a strong case that Newton in fact held the Arian position "well before 1675" (p. 315).

56 Whiston, *Athanasius Convicted of Forgery. In a Letter to Mr. Thirlby of Jesus-College in Cambridge* (London, 1712), p. 2. Thirlby argues, in reply, that "Legal Conviction" is simply inadequate. In such a grave matter, only absolute certainty, that is, "Full Proof," will serve; and it is not adequate to complain, as Whiston has, that Thirlby requires from Whiston "stronger *Proof* than the Nature of the Thing will admit of." Styan Thirlby, *A Defense of the Answer to Mr. Whiston's Suspicions, and an Answer to his Charge of Forgery Against St. Athanasius. In a Letter to Mr. Whiston* (London, 1713), p. 6.

57 Whiston, *Athanasius Convicted,* pp. 22ff.

58 Newton, Question: "Whether Anthanasius [& his friends] did not corrupt the records of the Council of Nice & Sardica" (brackets are in the original), in "Paradoxical questions concerning ye morals and actions of Athanasius and his followers," (holograph manuscript, leaf 7ᵛ, William Andrews Clark Memorial Library, UCLA).

59 Newton, Question: "Whether – Athanasius for shifting objectings taken from ye writings of Dionysius of Alexandria & from ye Council of Antioch collected against Paul of Samosat and for changing the ancient Doxology did not feign records," in "Paradoxical questions concerning . . . Athanasius," leaves 1ʳ, 2ʳ. Cf. Westfall's analysis of the Keynes and Yehuda manuscripts, *Never at Rest,* p. 314. Westfall states that not long after Newton discussed Whiston's *New Theory* with Whiston and secured Whiston's appointment as his substitute at Cambridge, "Whiston became the articulate spokesman for views virtually identical to Newton's" (ibid., p. 501).

60 Whiston, *Athanasius Convicted,* p. 1. Whiston adds that he is including the information that it was not originally his own idea that Athanasius was a forger so that Thirlby will not reject the evidence for this view out of hand, simply because it is presented by Whiston.

61 On the connection between Clarke's Arianism and Newtonian metaphysics, see Larry Stewart, "Samuel Clarke, Newtonianism, and the Factions of Post-Revolutionary England," *Journal of the History of Ideas* 42, no. 1 (Jan.–March 1981):53–72. Stewart shows that John Toland's argument against Newton's theory of passive, extended matter and void space was countered by Samuel Clarke's Boyle Lectures entitled *A Demonstration of the Being and Attributes of God* (London, 1705). Clarke argued against Toland's Cartesian concept of the infinite materiality of space by demonstrating the existence of a vacuum. And, "from this Clarke launched into the *a priori* argument based on the necessity of a universal self-existent Being whose attributes must be eternity, infinity, and unity" (Stewart, "Samuel Clarke," p. 56). Whiston claims that the heads of the colleges at Cambridge persecuted and expelled him "for the very same Christian Doctrines that the great Sir I. N. had discovered and embraced many Years before me; and for which Christian Doctrines, had He ventured as plainly and openly to publish them to the World as I thought myself oblig'd to do my own Discoveries, they must 30 or 40 Years ago have *Expell'd and Persecuted the Great Sir* Isaac Newton *also*" (Whiston, *Authentick Records,* 2:1080).

62 Eamon Duffy, 'Whiston's Affair': The Trials of a Primitive Christian, 1709–1714," *Journal of Ecclesiastical History* 27, no. 2 (April 1976), p. 140. Bishop Wake of Lincoln had urged Whiston to submit his writings to the Convocation. Whiston did and dedicated his *Historical Preface to Primitive Christianity* (London, 1710) "to the Most Reverend Reverend Thomas Lord Archbishop of Canterbury, President; and to The Right Reverend The Bishops of the same Province, His Graces Suffragans; and to the Reverend The Clergy of the Lower-House In Convocation Assembled." The Convocation deemed this dedication an insult.

63 G. V. Bennett, "Convocation of 1710," pp. 314–15.

64 Jonathan Swift, *Memoirs of Martinus Scriblerus,* ed. C. Kerby Miller (New Haven: Yale University Press, 1950), p. 285, note 13. Cited in Olson, "Tory–High Church Opposition to Science," p. 12.

65 Sir Richard Cox to Edward Southwell, Dec. 23, 1714, quoted in C. J. Rawson, "Parnell on Whiston," *Bibliographical Society of American Papers* 57 (1963):91–2. Parnell's poem later was published in *The Miscellanies: The Last Volume* (London, 1727), pp. 172–3.
66 Alexander Pope, "God's Revenge against Punning," p. 1.
67 *Examiner,* IV, 41 (October 26, 1712), and V, 12 (January 8, 1713), in Blanchard, *Englishman.*
68 Whiston, *Memoirs,* 1:208.
69 Westfall, *Never at Rest,* p. 651.
70 Olson, "Tory–High Church Opposition to Science," p. 21.
71 See Henry Horwitz, *Revolution Politicks: The Career of Daniel Finch, Second Earl of Nottingham, 1647–1730* (Cambridge: Cambridge University Press, 1968), p. 256.
72 Whiston, *Historical Memoirs of the Life of Dr. Samuel Clarke. Being a Supplement to Dr. Sykes's and Bishop Hoadley's Accounts. Including certain Memoirs of several of Dr. Clarke's Friends* (London, 1730), pp. 76–80. Characteristically, Whiston waited until Clarke's death before telling his own side of the tale.
73 See Robert E. Sullivan, *John Toland and the Deist Controversy: A Study in Adaptations* (Cambridge, Mass.: Harvard University Press, 1982), p. 136. See also Geoffrey Holmes, "Science, Reason, and Religion in the Age of Newton," pp. 169–70.
74 Whiston, *Memoirs,* 1:178. In 1746 Whiston published Bishop Thomas Sherlock's sermon condemning the 1745 rebellion with an appendix of his own explaining how the rebellion fitted into the divine scheme. Whiston had already shown in *An Essay on the Revelation of St. John, So far as Concerns the Past and Present Times* (London, 1706) that the Protestants were in no danger of succumbing to Popery (pp. 309–10). Now, providing only that the Protestants are not obstinate "against farther Reformations, to a prodigious Degree, we may justly depend at this . . . *last* Crisis, by the same Interposition of Providence for our Preservation." See *A Sermon Preached at the Cathedral Church of Salisbury, October the 6th, 1745. On the Occasion of the Rebellion in Scotland. By the Right Reverend Thomas Lord Bishop of Salisbury. Published at the Request of the Mayor and Corporation: But Re-published by Mr. Whiston, with some Additions, 1745–6* (London, 1746), p. 16. For a discussion of how the London earthquakes fitted into Whiston's interpretation of God's providential plan revealed in prophecy, see Chapter 5. One reason for Whiston's great interest in eclipses was their connection with the fulfillment of prophecies. He claims that "the Grand *intermediate Breaches* in every one of these four Monarchies [prophesied in the Book of Daniel and symbolized by Daniel's vision of metallic men – golden Babylon, silver Persia, bronze Greece, and iron Rome – Dan. 2] was immediately preceded by great Eclipses of the Moon" (Whiston, *Memoirs of Clarke,* p. 150). Whiston interprets the great solar eclipse of 1736 as a "divine signal for over-bearing persecution in the ten idolatrous and persecuting Kingdoms which arose in the fifth century in the *Roman* empire, the *Britains* and the *Saxons*" (Whiston, *Memoirs,* 1:205–6; cf. Whiston, *Essay on the Revelation,* pp. 323–4, and the pamphlet entitled *Astronomical Year* [London, 1737]).
75 Whiston, *Memoirs,* 1:175–6. The reply from Prince Eugene is cited in DNB, s.v. "Whiston, William." Whiston discussed this interpretation of the Turkish defeat with his friend, the great Whig general, James Stanhope, shortly after Stanhope

told him the news. Stanhope was in accord with Whiston about this particular point, as well as with Whiston's general "Scheme of Scripture Prophecies." See Whiston, *The Literal Accomplishment of Scripture Prophecies. Being a full Answer to a late Discourse, Of the Grounds and Reasons of the Christian Religion* (London, 1724), p. 97.

76 Whiston, *Memoirs,* 1:31.

77 Westfall, *Never at Rest,* pp. 325, 816–17.

78 Whiston, *Memoirs,* 1:35–6, 250.

79 Enough has been said of Whiston's literalistic method. Regarding Newton's literal rules for the interpretation of millennial prophecy, see Westfall, *Never at Rest,* p. 326.

80 Whiston, *Essay on the Revelation,* pp. 112–13.

81 Ibid., p. 245.

82 Whiston, "Corrections and Improvements to the *Essay on the Revelation of St. John,*" in *Literal Accomplishment of Scripture Prophecies,* p. 102.

83 Ibid., p. 110.

84 Ibid., p. 103.

85 Ibid., p. 116.

86 Whiston, "Of the Restoration of the Jews," in *Sermons and Essays,* pp. 224–5.

87 Ibid., pp. 233–4.

88 Westfall, *Never at Rest,* pp. 321–5.

89 Whiston, *Memoirs,* 1:152–3. On Newton's views on interpreting prophecies, see Westfall, *Never at Rest,* pp. 321–5. Newton kept his rage silent.

90 Jacob, *Newtonians and the English Revolution,* p. 133.

5. DEISM AND DIVINE PROVIDENCE IN WHISTON AND NEWTON

1 Frank E. Manuel, *Isaac Newton, Historian* (Cambridge, Mass.: Harvard University Press, 1963), p. 143.

2 Richard S. Westfall, "Isaac Newton's 'Theologiae Gentilis Origines Philosophicae,'" in *The Secular Mind: Transformations of Faith in Modern Europe: Essays Presented to Franklin L. Baumer, Randolph W. Townsend Professor of History, Yale University,* ed. W. Warren Wagar (New York: Holmes & Meier, 1982), pp. 30–1.

3 Manuel, *Newton, Historian,* pp. 170–7.

4 Henry Guerlac, "Theological Voluntarism and Biblical Analogies in Newton's Physical Thought," *Journal of the History of Ideas* 44, no. 2 (April–June 1983): 227–8.

5 For an examination of this development among various members of the Royal Society, see James E. Force, "Hume and the Relation of Science to Religion among Certain Members of the Royal Society," *Journal of the History of Ideas* 45, no. 4 (Oct.–Dec. 1984).

6 John Wilkins, *Of the Principles and Duties of Natural Religion* (London, 1683), p. 130.

7 Newton conducted a correspondence with Bentley beginning five days after Bentley's final lecture. Bentley's use of Newton's law of gravity to demonstrate the existence of a generally provident creator-architect occurs in the last three published versions of these lectures, which are separately entitled "A Confutation of Atheism from the Origin and Frame of the World," in *A Defence of Natural and Revealed Religion: Being a Collection of the Sermons Preached at the Lecture*

founded by the Honourable Robert Boyle, Esq; (From the Year 1691 to the Year 1732), ed. Sampson Letsome and John Nicholl (London, 1739), 3 vols., 1:1–11. In his letters to Bentley, Newton indicates that he is delighted with Bentley's application of his "systeme" to support a belief in a generally provident deity skilled in geometry and physics. Newton himself equivocates about whether the direct cause of gravity is material or immaterial, but Bentley, along with most of Newton's other disciples, reiterates that the cause of gravity is immaterial, divine power. There is good evidence that Newton participated with the executors of Boyle's will, especially Pepys (at one time president of the Royal Society) and Evelyn, in handpicking Bentley as the first Boyle lecturer. See H. Guerlac and M. C. Jacob, "Bentley, Newton, and Providence," *Journal of the History of Ideas* 30 (July–Sept. 1969): 318.

8 *The Correspondence of Isaac Newton*, 7 vols. (Cambridge: Cambridge University Press, 1959–77), vol. 3, ed. H. W. Turnbull, 336.

9 See *A Collection of Papers, Which passed between . . . Mr. Leibnitz and Dr. Clarke* (London 1717), p. 15, where Clarke observes that " 'tis not a *diminution*, but the true *Glory* of his Workmanship, that *nothing* is done without his *continual Government and inspection*."

10 Whiston, *A New Theory of the Earth*, 2nd ed. (London, 1708), p. 284.

11 Ibid.

12 *Sir Isaac Newton's Theological Manuscripts*, ed. H. McLachlan (Liverpool: at the University Press, 1930), p. 17. Like Whiston, Newton believed that miracles ceased early in the church's history. See Newton to Locke, February 16, 1691–92, in *Correspondence of Newton*, 3:195.

13 See Newton's correspondence with Thomas Burnet on this point in Turnbull, *Correspondence of Newton*, 2:331. In the "Introductory Discourse" to his *New Theory*, Whiston explains that "the Mosaick Creation is not a Nice and Philosophical account of the Origin of All Things; but an Historical and True Representation of the formation of our single Earth out of a confus'd Chaos, and of the successive and visible changes thereof each day, till it became the habitation of Mankind" (p. 3). For Whiston and for Newton, the account of the secondary mechanisms of nature that lie behind Moses' adaptation for the vulgar constitutes a "Nice and Philosophical account of the Origin of All Things." The chief natural mechanism is that of a comet and, on the view of Whiston and Newton, its operation in accord with natural law to effect "the formation of our single Earth out of a confus'd Chaos" is itself a miraculous reflection of God's continuous special providence. Newton very much approved Whiston's book.

14 John Keill, *An Examination of the Reflections on the Theory of the Earth* (Oxford, 1699).

15 Whiston believed in the miracles of Jesus and that the Apostles continued to have specially provident power to contravene the laws of nature in such supernatural, miraculous acts as the curing of disease by the laying on of hands. See *Mr. Whiston's Account of the Exact Time When Miraculous Gifts Ceas'd in the Church* (London, 1728), p. 7. Though he believed in the power of the Apostles to effect miracles in their time and that such power continued in the first three and a half centuries among Christians, Whiston, like Wilkins, was highly suspicious of reports of modern-day miracles. Even so, he recounts the story of one John Duncalf, who stole a Bible. When challenged, Duncalf denied he was the

thief and swore that if he had stolen the Bible, he desired his hands to "rot off"–which, supposedly, happened. Whiston recommends that the exact narrative of this event "ought, in this sceptical age, to be reprinted, and recommended to all, who either deny, or doubt of the *interposition of a particular divine providence*" (emphasis added). See Whiston, *Memoirs of the Life and Writings of Mr. William Whiston . . .*, 2nd ed., 2 vols. (London, 1753), 1:6. Unless otherwise noted, all citations are to this edition. Charles Webster has recently shown that Whiston's defense of the real existence of demonic magic also derives from his view that miracles occurred in the past and still occur in the present as the result of the interposition of wicked, but specially provident, demonic agents. Webster states, "To Whiston, demons were just as necessary for the concept of Providence as eclipses, comets, northern lights, meteors, and earthquakes. Any purely naturalistic explanation of these phenomena would detract from the idea of the extraordinary powers of God, and thereby cast doubt on the entire scheme of universal history. In particular the climactic phase of the eschatological sequence would involve the particularly active intervention of the Devil, as the elect wrestled with agents of Antichrist." Charles Webster, *From Paracelsus to Newton. Magic and the Making of Modern Science* (Cambridge: Cambridge University Press, 1982), p. 98. Whiston's belief that special providence could be exercised by both bad agents (demons) and good agents (angels) accounts for his remark to John Lacy, one of the converts to the evangelical and millennial French Prophets who preached in London beginning in 1706. Whiston states, "About the year 1713, I held a conference at my house . . . with Mr. *Lacy,* and several other of those prophets; wherein I gave my reasons why, upon supposition of their agitations and impulses being *supernatural,* I thought they were *evil* and not *good* spirits that were the authors of those agitations and impulses" (Whiston, *Memoirs,* 2:119).

16 Whiston, *The Accomplishment of Scripture Prophecy* (London, 1708), p. 5.

17 *A Collection of Papers which passed between . . . Mr. Leibnitz and Dr. Clarke,* p. 15. In general, this entire series of correspondence is yet one more attempt by an eminent Newtonian, working, in all probability, in collaboration with Newton himself, to retain the concept of a God who is specially provident, and not just an absconding creator-designer. As Koyré points out, it is "utterly inconceivable that Clarke should accept the role of philosophical spokesman (and defender) of Newton without being entrusted by the latter to do it." See Alexandre Koyré, *From the Closed World to the Infinite Universe* (Baltimore: Johns Hopkins University Press, 1976), p. 301.

18 Samuel Clarke, *A Discourse Concerning the Unchangeable Obligations of Natural Religion, and the Truth and Certainty of the Christian Revelation, Being Eight Sermons Preached at the Cathedral Church of St. Paul in the Year 1705 in A Defence of Natural and Revealed Religion,* 2 vols., 2:165.

19 James E. Force, "Hume and Johnson on Prophecy: The Context of Hume's Essay 'Of Miracles,'" *Journal of the History of Ideas* 40, no. 3 (July–Sept. 1982):463–75.

20 Locke writes that the miracles of Jesus are significant only because they are predicted in biblical prophecy. See *The Reasonableness of Christianity as Delivered in the Scriptures* (London, 1695), p. 55. Newton agrees that "the events of things predicted many ages before . . . will then be a convincing argument that

the world is governed by *providence*" (*Observations upon the Prophecies of Daniel and the Apocalypse of St. John* [London, 1733], p. 252).

21 Newton, *Observations upon the Prophecies*, p. 14.

22 On the basis of Guerlac's and Jacob's picture of Newton's influence upon the executors of Boyle's will in picking Bentley as the inaugural lecturer ("Bentley, Newton, and Providence"), it seems probable, given Newton's evident interest in the selection of Whiston, that Newton helped get Whiston appointed to give the Boyle Lectures. Newton had already furthered Whiston's career by securing for Whiston Newton's own chair at Cambridge when he left the university in 1703. However, there is no direct evidence that Newton intervened to help get Whiston appointed as Boyle lecturer. Whiston does state that his topic for the series of lectures was originally suggested by Newton himself. In the midst of a long passage in his *Memoirs* concerning the "ludicrous" skepticism engendered by the allegorical or figurative method of interpreting Scripture (in contrast to Whiston's and Newton's "literalistic" hermeneutic method), Whiston complains that the allegorical method was so entrenched that even such "learned men as bishop *Chandler*, and *Dr. Clarke*, made use (of it): till I, upon Sir Isaac Newton's *original suggestion*, shewed them the contrary" (p. 98). The work in which Whiston attacked Clarke and Chandler appeared in 1725, but this work is merely an appendix to the 1708 Boyle Lectures. Thus it seems probable that Whiston's Boyle Lecture series topic was first suggested by Newton. If this is the case, Guerlac's and Jacob's contention that Newton used the public platform to promulgate by proxy his views on controversial religious questions is strengthened. See also Whiston's *Supplement to the Literal Accomplishment of Scripture Prophecies* (London, 1725), pp. 4–5.

23 In his years as president of the Royal Society from 1701 to 1727, Newton was a total dictator. He endorsed only those who agreed with his views. Eventually Newton even had a quarrel with Whiston and prevented his attaining membership in the Society. In 1720, Newton's close friends and advisers, Sir Hans Sloane and Edmond Halley, proposed Whiston for membership, but Newton blocked Whiston's election by threatening to resign as president, an action prompted by the conflict between two strong wills. Whiston writes that "if the reader desires to know the reason of Sir *Isaac Newton's* unwillingness to have me a member, he must take notice, that as his making me first his deputy, and giving me the full profits of the place brought me to the heads of the colleges in Cambridge, made me his successor; so did I enjoy a large portion of his favour for twenty years together. But he then perceiving that I could not do as his other darling friends did, that is, learn of him, without contradicting him, he could not, in his old age, bear such contradiction; and so he was afraid of me the last thirteen years of his life. . . . He was of the most fearful, cautious, and suspicious temper, that I ever knew" (*Memoirs*, 1:250–1). Although allowed to present his papers before the Society whenever he pleased, Whiston was never admitted as a Fellow, owing to Newton's personal enmity. Nevertheless, Whiston's views on the importance of prophecy for demonstrating divine special providence were repeated by so many Fellows of the Royal Society that I feel justified in using his texts to illustrate the official apologetic position of the "first" Royal Society. For objective accounts of Newton in his later years as president of the Royal Society, see Christopher Hill, *Change and Continuity in 17th Century England* (London: Weidenfeld & Nicolson, 1974), Chapter 12, and Frank E. Manuel, *Freedom*

from History and Other Essays (New York: New York University Press, 1972), Chapter 7.

24 William Stukeley, *Family Memoirs of the Rev. William Stukeley, M.D.* (London, 1880), vol. 1, "The Commonplace Book, 1720," p. 100 (emphasis added). Folkes was vice-president of the Royal Society while Newton was president and chaired many meetings in Newton's absence. His influence within the society predated his own election to the presidency by a considerable period. The "heathen" Will Jones mentioned by Stukeley was the man who taught astronomy and mathematics to George, Second Earl of Macclesfield, at the earl's residence, Shirburn Castle in Oxfordshire. This is significant because Macclesfield was highly active in the council during Folkes's tenure as president and succeeded Folkes in that post, serving from 1752 to 1764. See Lyons, *The Royal Society 1660–1940: A History of Its Administration Under Its Charters* (New York: Greenwood Press, 1968), p. 181.

25 Whiston, *Memoirs,* 1st ed., 2 vols. (London, 1749), 1:333.

26 The essay was published in Swift's and Pope's *Miscellanies. The Third Volume* (London, 1732), pp. 255–76.

27 Ibid., p. 257.

28 Ibid., p. 260.

29 Ibid., p. 262.

30 Ibid., p. 263.

31 Ibid., p. 272.

32 Ibid.

33 Ibid., pp. 274–5.

34 Ibid., pp. 275–6.

35 Whiston, *An Account of a Surprizing Meteor, Seen in the Air March the 6th, 1715/16, at Night* (London, 1716), pp. 75–7.

36 Alexander Pope, "God's Revenge against Punning," in Swift and Pope, *Miscellanies: The Third Volume,* p. 53.

37 Jonathan Swift, *Dean Swift's True . . . Copy of that . . . Surprizing Prophecy Written by St. Patrick* (London, 1740), p. 7. For a modern interpretation of Whiston as a "learned fool" and for a discussion of Gay's authorship of "A True and Faithful Narrative," see James M. Osborn, " 'That on Whiston,' " *Papers of the Bibliographical Society of America* 56 (1962):73–8.

38 See G. S. Rousseau, "The London Earthquakes of 1750," *Journal of World History* 11, no. 3 (1968):436–51.

39 William Stukeley, *The Philosophy of Earthquakes, Natural and Religious. Or an Inquiry into their Cause, and their Purpose* (London, 1750), p. 5. This pamphlet, dedicated to the detested Martin Folkes, president of the Royal Society, contains two papers read to the Royal Society on March 15 and March 22, 1750, as well as a sermon Stukeley preached in his own church, St. George's, in Queen Square.

40 Ibid., p. 20.

41 Ibid.

42 Ibid., p. 46.

43 Ibid., p. 42. Though he calls this effect a "miraculous" one and states that it could not be due to natural causes, Stukeley seems to mean that it is a "prodigious" effect produced by a supreme artisan. He apparently envisions God, as an updated version of Zeus, seated at a master control board, spinning dials and

throwing switches that hurl down lightning bolts, which, when they strike home, set off earthquakes precisely calculated to shake but not destroy. There seems to be nothing "miraculous," in Hume's sense, in this view.

44 Ibid.

45 Whiston, *Memoirs,* 2:14.

46 Ibid., pp. 59–69.

47 Ibid., pp. 135–9.

48 Whiston, *Daily Advertiser,* March 14, 1750. Reprinted in his *Memoirs,* 2:215. The source for the number of people attending Whiston's lectures is T. D. Kendrick, *The Lisbon Earthquake* (London: Methuen, 1956), Chapter 1.

49 Thomas Secker, "Preached on the Occasion of the Earthquake, March 8, 1749–50. Psalm ii. 11. *Serve the Lord with Fear and rejoice unto him with Reverence,*" in *Fourteen Sermons Preached on Several Occasions* (London, 1766), Sermon VIII, p. 205.

50 Ibid. What is "totally out of Sight" is the time of the event, not the knowledge that the event will happen.

51 *General Evening Post,* April 17–19, 1750. It is possible that Whiston was "Publicus," since he quotes this entire letter in his *Memoirs,* 2:209–13, and says that he wishes the citizenry of London to accept it "as my own Address to them, upon this solemn Occasion." One hundred thousand is undoubtedly a highly exaggerated number.

52 Walpole to Horace Mann, April 4, 1750, in *Letters of Walpole,* ed. Paget Toynbee, 19 vols. (Oxford: Clarendon Press, 1903–25), 2:440.

53 Samuel Chandler, elected a Fellow of the Royal Society in 1754, published *The Scripture Account of the Cause and Intention of Earthquakes, in a Sermon preached at the Old-Jury, March 11,* [1750], *On Occasion of the Two Shocks of an Earthquake, the First on February 8, the Other on March 8* (London, 1750); for works by two other Fellows, see Theophilus Lobb, *Sacred Declarations* (London, 1750), and Robert Pickering, *An Address to those who have either retired, or intend to leave the town, under . . . apprehension of . . . another earthquake. Being the subject of a sermon* [on Ps. 139:7, 9–10] (London, 1750).

54 Stephen Hales, "Some Considerations on the Causes of Earthquakes," *Philosophical Transactions of the Royal Society of London* 497 (1750):669.

55 Hume to Clephane, April 1750, in *The Letters of David Hume,* ed. J. Y. T. Greig, 2 vols. (Oxford: Clarendon Press, 1932), 1:141.

56 Walpole to Mann, April 2, 1750, in *Letters of Walpole,* 2:435.

57 [Paul Whitehead?], *A Full and True Account of the Dreadful and Melancholy Earthquake which happened between Twelve and One o'clock in the Morning, on Thursday the Fifth Instant. With an Exact List of such Persons as have hitherto been found in the Rubbish. In a Letter from a Gentleman in Town, to his Friend in the Country* (London, 1750), p. 4.

58 Ibid., p. 7. This particular kind of satire, which details the results of some imaginary catastrophe predicted prophetically, had already been directed against Whiston, probably around 1729 or 1730, by John Gay, according to Osborne, " 'That on Whiston,' ":73–8.

59 *The Military Prophet's Apology; or, Probable Reasons for deferring the Earthquake; in a Letter to Dr. Middleton* (London, 1750).

60 Voltaire, *The Metaphysics of Sir Isaac Newton,* trans. David Erskine Baker (London, 1747), p. 3.

61 Whiston, "Reason and Philosophy no Enemies of Faith," in *Sermons and Essays upon Several Subjects* (London, 1709), p. 210.
62 Richard S. Westfall, "Newton's 'Theologiae . . . Philosophicae,' " p. 16.
63 Ibid.
64 Manuel, *Newton, Historian*, p. 49.
65 Westfall, "Newton's 'Theologiae . . . Philosophicae,' " p. 22.
66 Ibid., p. 23.
67 Ibid., p. 24.
68 Ibid., p. 25.
69 McLachlan, *Newton's Theological Manuscripts*, p. 3.
70 Arthur Young, *An Historical Dissertation on Idolatrous Corruptions in Religion*, 2 vols. (London, 1734), 2:269.
71 Whiston, *A Collection of Authentick Records Belonging to the Old and New Testament*, 2 vols. (London, 1728), 2:1077.
72 Ibid., p. 1079.
73 Whiston, *Memoirs*, 1st ed. (1749), 2 vols., 1:39–40.
74 Whiston, *Authentick Records*, pp. 962–4.
75 Ibid., pp. 1055–6.
76 Manuel has discovered manuscript sources in which Newton claims that the Scriptures "are by far the oldest records now extant" and that because of the more ancient literary tradition of Moses and the Israelites, the reputation of the sacred histories is "much above [that of] the histories of Manetho, Berossus, Ctesias, Herodotus, Megasthenes, Diodorus, Sanconiatho, or any other heathen histories" (*Newton, Historian*, p. 58). Still, in the light of his deistic history of the development of the Gentile nations and their theology, Newton in practice occasionally interprets the Mosaic history in the light of the pagan accounts. Because the Hebrew Scriptures are the oldest in general does not mean that they contain no errors.
77 Whiston, *Authentick Records*, p. 991.
78 Newton, *Sir Isaac Newton's Mathematical Principles of Natural Philosophy and His System of the World*, trans. Andrew Motte, rev. Florian Cajori (Berkeley: University of California Press, 1934), p. 490. Newton calculates the precession rate at "about 50″ yearly."
79 Whiston, *Authentick Records*, p. 1010.
80 Manuel, *Newton, Historian*, p. 81.
81 Whiston, *Authentick Records*, p. 1011.
82 Manuel, *Newton, Historian*, p. 174.
83 Whiston, *Authentick Records*, p. 963.
84 Ibid., p. 1056.
85 Hume, "Of a Particular Providence and a Future State," Section XI of *An Enquiry Concerning Human Understanding*, in *Essays and Treatises on Several Subjects*, 2 vols. (London, 1777), 2:104.
86 Ibid., p. 105.
87 In the face of such complicated subjects, Hume inquires, "Why torture your brain to justify the course of nature upon suppositions, which for aught you know, may be entirely imaginary, and of which there are to be found no traces in the course of nature?" (ibid., p. 107).
88 Ibid.

89 Hume, "Of Miracles," Section X of *Enquiry Concerning the Human Understanding*, pp. 87–88.
90 Ibid., p. 90.
91 Ibid., p. 91.
92 Ibid., p. 100.
93 Ibid., p. 101.
94 All of the definitions s.vv. "prophecy" and "to prophesy" given in Samuel Johnson, *A Dictionary of the English Language: in which the Words are deduced from their Originals, and Illustrated in their Different Significations by Examples from the best Writers,* 2nd ed., 2 vols. (London, 1756), show this identity. Prophecy is "a declaration of something to come; prediction." *To prophesy* (v.a.) is "1. To predict; to foretell; to prognosticate. 2. To foreshow." *To prophesy* (v.n.) is "to utter predictions." Conversely, *to predict* is "to fortell; to foreshow," and *prediction* is simply "prophecy; declaration of something future."
95 St. Augustine, *Confessions,* Book VII, Chapter 6, trans. R. S. Pine-Coffin (Harmondsworth, England: Penguin Books, 1981), p. 142.
96 Dante Alighieri, *The Inferno,* Canto XX, trans. John Ciardi (New York: New American Library [Mentor Books], 1954), pp. 174–81.
97 Johnson, *Dictionary,* s.v. "prophet." Johnson gives two definitions. The first and primary definition is "one who foretells future events; a predicter." Under this head are scientists, astrologers, and soothsayers. The second definition, just quoted in the text, connects the ability to predict the future with God's power to grant this superhuman ability. This is the most significant sense of the term for Hume's argument.
98 Whiston, *Supplement to the Literal Accomplishment of Scripture Prophecies,* pp. 5–6.
99 Ibid., p. 8.
100 Ibid., pp. 1–2.
101 Whiston, *Accomplishment of Scripture Prophecy,* p. 1. In addition to this work and the work cited in note 99, Whiston has lists of completed "historical" prophecies, as well as a sprinkling of unfulfilled "future" prophecies, in his *Literal Accomplishment of Scripture Prophecies* (London, 1724) and in the second volume of the second edition of his *Memoirs,* which bears the subtitle *To which are added his Lectures on the late Remarkable Meteors and Earthquakes, and on the Future Restoration of the Jews.*
102 Newton, *Observations Upon the Prophecies,* pp. 252–3.
103 Antony Flew, in a recent article ("Parapsychology: Science or Pseudo-Science?" *Pacific Philosophical Quarterly* [1980]:100–14), explains why parapsychology ought to be classified, for now, as a pseudoscience and how Hume's argument against miracles applies to the data of parapsychological experimentation. Parapsychology is defined as the study of "psi-phenomena," which includes "psi-kappa phenomena" (psychokinesis – movement by the mind) and "psi-gamma phenomena" (clairvoyance – mind communicating to mind – and telepathy – matter communicating to mind). The most important aspect of psi-gamma phenomena, in the context of the present essay, are precognitive psi-gamma phenomena. The subject of an experiment involving precognitive psi-gamma phenomena will receive or produce or derive information from outside him- or herself (through clairvoyance or telepathy) normally available only in the future. Such paranormal precognition corresponds to my category of astrologers who

derive their advance information "from the stars." Flew categorizes parapsychology as a pseudoscience because of "impressively disillusioning instances of fraud and self-deception" and because there is no "repeatable demonstration of the reality of such phenomena." Both objections apply to astrology as well. Hume's argument applies to precognitive psi-gamma phenomena, because "any piece of work claiming to show that psi-phenomena have occurred is in effect a miracle story." Flew points out that "this means that we have to interpret and assess the available evidence in light of all we know, or think we know, about what is probable or improbable, possible or impossible. But now . . . psi-phenomena are implicitly defined in terms of the violation of some of our most fundamental and best evidenced notions of contingent impossibility. So, even before any Humian allowance is made for the special corruptions afflicting this particular field, it would seem that our historical verdict will have to be, at best, an appropriately Scottish, and damping: 'Not proven' " (p. 108).

The effect of Flew's analysis is to show that stories of paranormal precognition are implicit miracle stories because they transgress our experience of natural law. The predictive utterances of soothsayers, of those experiencing apparent paranormal precognition, and of Old Testament prophets are all, in the Humian sense, miraculous and hence unbelievable. My point is that Hume himself understands his argument to apply at least to Old Testament prophetic predictions. I am much indebted to Professor Flew for bringing this article to my attention.

104 David Hume, *Dialogues Concerning Natural Religion,* ed. Norman Kemp Smith (Indianapolis: Bobbs-Merrill, 1947), Part 1, pp. 134–5.

105 Woolston actually does connect his attack on miracles with a brief attack on prophecies: "To prove the Messiahship of the Holy Jesus from his Miracles, our *Divines* urge the Prophecys of the Old Testament, such as that of Isaiah, c.xxxv. v. 5, 6 . . . and say that these Prophecys were accurately fulfill'd by our *Jesus* in the several Cures of Blindness, Deafness, Lameness, and Dumbness, which he often perform'd upon one and another; and, in asmuch as our Saviour seems to appeal to such Prophecys, do conclude that his Accomplishment of them to be no less than a Demonstration, that he was the true *Messiah,* that great Prophet, who was to come into the World. To which I answer, *First,* that the Accomplishment of Prophecys that can neither be given forth by human Foresight, nor fulfill'd in a Counterfeit, are good Proofs of *Jesus' Messiahship;* But then, what shall we say if others besides *Jesus* should do the like Cures and Miracles? It is said of *Antichrist,* and I believe it, that he will not only do all the Miracles, that Jesus did, but will appeal to the like Prophecys too. How then we are to distinguish the true Christ from the false Christ by Miracles and Prophecys in this Case, is the Question, which I leave with our Divines to consider of an Answer to" (*A Discourse on the Miracles of our Saviour, In View of the Present Controversy between Infidels and Apostates* [London, 1727], pp. 15–16). Woolston never attacks the design argument, however.

106 Hume, *The History of England, from the Invasion of Julius Caesar to the Revolution in 1688,* 8 vols. (London, 1782), 6:196–7.

107 Percy Bysshe Shelley, *Queen Mab, a Philosophical Poem* (London, 1813), pp. 203–4.

108 Cited in Constance A. Lubbock, ed., *The Herschel Chronicle: The Life-Story of William Herschel and His Sister Caroline* (Cambridge: Cambridge University Press, 1933), p. 310.

109 Shelley, *Queen Mab,* p. 184. Shelley refers to "*Sir W. Drummond's Academical Questions*" (London, 1805) and states that Drummond seems to construe the inevitable step toward atheism to which Newtonian mechanical philosophy seems to lead as "a presumption of falsehood" against that system. But Shelley believes it is more honest "to admit a deduction from facts than an hypothesis incapable of proof, although it might militate with the obstinate preconceptions of the mob."

INDEX

Printed in the United States
By Bookmasters